HANDBOOK OF
MODEL LETTERS AND MEMOS
FOR ENGINEERS, SCIENTISTS
AND TECHNICAL PROFESSIONALS

George E. Parker

PRENTICE HALL
Englewood Cliffs, New Jersey 07632

Prentice-Hall International (UK) Limited, *London*
Prentice-Hall of Australia Pty. Limited, *Sydney*
Prentice-Hall Canada, Inc., *Toronto*
Prentice-Hall Hispanoamericana, S.A., *Mexico*
Prentice-Hall of India Private Limited, *New Delhi*
Prentice-Hall of Japan, Inc., *Tokyo*
Simon & Schuster Asia Pte. Ltd., *Singapore*
Editora Prentice-Hall do Brasil, Ltda., *Rio de Janeiro*

© 1988 *by*

PRENTICE-HALL, Inc.

Englewood Cliffs, NJ

10 9 8 7 6 5 4 3 2 1

To Irma

Library of Congress Cataloging-in-Publication Data

ISBN   0-13-380429-1   NBZI

PRENTICE HALL
BUSINESS & PROFESSIONAL DIVISION
A division of Simon & Schuster
Englewood Cliffs, New Jersey 07632

Printed in the United States of America

ACKNOWLEDGMENT

Working with hundreds of authors over the years has taught me, among other things, that a book is never the product of a single individual. This one is no exception. My sincere appreciation is extended to all those who contributed so much to the work that follows. They include Martha Cresci, Gerald Tomlinson, Harold E. Meyer, J. A. Van Duyn, George Z. Kuredjian, Barbara Palumbo, Sally Ann Bailey, Michael Alley, and Donald Helgeson. Special thanks also, to a valued friend and ally, Jo Anne Kern.

A WORD FROM THE AUTHOR ON THE PRACTICAL VALUE THIS BOOK OFFERS . . .

At the core of any technical profession lies the art of communicating data effectively. And within this intricate, challenging world, skill in *written* communication becomes particularly important. It is precisely in this area—conveying one's documented thoughts in a clear and compelling way—that this daily reference book will provide invaluable assistance.

Clarity of thought can be reflected by what we say, but it takes on an added, more lasting dimension when we express the same degree of clarity in our writing. Few would disagree with the notion that written documents not only record technical progress; they contribute immeasurably to it. However, as with other factors that affect our performance, more is not necessarily better when communicating to others. Excessive words cloud the issue just as much as a presentation that omits key factors. To some, combining clarity with brevity is a mysterious art. To others, the impact of a powerfully written statement may not even be realized immediately. Consider the following, for example.

Abraham Lincoln shortly after the remarks he offered at Gettysburg referred to his address as a "flat failure." Many who attended the ceremony agreed with him. He had been preceded on the platform by Edward Everett—statesman, scholar, and university president. Everett delivered a 117-minute address, in which he used all his impressive gifts of oratory and elicited great praise. Following this dramatic performance, Lincoln was introduced. He spoke for 2 minutes. The next day Everett wrote Lincoln, "I wish I could flatter myself that I had come as near to the central idea of the occasion in two hours as you did in two minutes." Consistent with this point, you will note the great majority of letters and memos that follow do not exceed one page in length.

Professionals active in any branch of technology are trained to place a high value on precision. They know that precision does not always come easily. It takes time. And that is one of the principal ways in which this book will be useful. It would be presumptuous to build this reference work on ways to "create" more effective written communications. Engineers and others in technical fields frequently write excellent letters and memos. The purpose here is to help you save valuable time by providing tested, useful models of written communications, all based on realistic situations. Many can be used verbatim; others can quickly be adapted to fit your particular situation. In addition, each letter offers alternative approaches that can further simplify adaptation.

Keep in mind when you use the letters in this book, and when you write your own, that the best modern letter writing requires a tone that is courteous and straight to the point. There is no place for stilted phrases or archaic forms. For the most part, the message should be written as you would tell it, editing for brevity, and designing the physical appearance to help get your message across. (See Chapter 10 for useful guidelines on the physical arrangement of letters.)

True, some well-written letters are especially graceful, some may even have a touch of humor, but generally, good letters need not possess these characteristics. There is no point in having the reader exclaim over your literary skill, or stand in awe of your vocabulary, if he or she ends up wondering what on earth you are talking about—and then does the wrong thing in response.

This daily reference work offers an extensive, carefully selected assortment of letters and memoranda that will not only save you time, it will help you secure the tangible results you want to achieve. Note also that many letters and memos fall neatly into more than one category. Thus, a letter in Chapter 4 might easily be adapted to secure technical information you need, or you might find a more suitable approach outlined in Chapter 1, dealing with "Request" letters.

To broaden and increase the book's value, a wide range of situations has been covered. There are letters and memos of appreciation, reprimand, congratulations, evaluation, and others dealing with such practical matters as technical reports, systems analysis, design flaws, complaints, new ventures, product opportunities, testing and evaluation, career planning, and much, much more.

You will also note the models have been drawn from realistic situations that most of us face on a daily basis. Many are designed to be productive in ways that allow results to be measured, not merely assumed. Productivity is just as important in letter writing as it is in the refinement of existing products or the development of new products. Thus, the models are geared to specific needs, not theoretical assumptions that exist only in the mind of the composer. We deal with the world of reality, and the real people who inhabit that world.

For example, Matthew Broderick was on the engineering staff of a leading electronics firm. They had a contract to deliver instruments to the Air Force by a specific date. However, one of the firm's prime suppliers questioned the use of certain parts that were crucial to the fulfillment of the contract. Initially, Broderick developed a three-page letter to the supplier, but discarded it because it did not put the primary message in sufficiently sharp focus. Finally, he called the supplier and followed up by sending an adapted version of the three-paragraph letter that appears in Chapter 5. The resulting agreement on revised specifications, in addition to the revenue from this assignment, helped pave the way for additional contracts.

Alan Gabacia was faced with a variation of this problem. His computer equipment firm had a target date for delivery of a new memory system to an important client. Because a design refinement in the system became more complex than he anticipated, he realized he needed additional time. The letter he used (see Letter 1-1) earned that additional time, while enabling him to underscore the benefits to be gained by the delay. Result: an extension was granted, and the customer's awareness of the new system's potential value was reinforced. While these examples illustrate the book's practical nature, we should also emphasize that the content and general tone used throughout will respond to particular needs in ways that are courteous, thoughtful, and professional.

As an added feature of this time-saving reference work, you will find two "bonus" chapters. First, a master checklist (Chapter 10) provides all the essentials of a successful letter, and, second, a comprehensive listing of proper address forms (Chapter 11) covers a broad range of officials in federal, state, and local government, including all branches of the armed forces.

In short, the *Handbook of Model Letters and Memos for Engineers, Scientists and Technical Professionals* is just what the title promises, a comprehensive and useful collection of tested, time-saving correspondence you can quickly and easily adapt to your particular needs. Most of all, it is a book that, right from the beginning, will help you write letters and memos that *get results*.

George E. Parker

HOW TO USE THIS BOOK

Mark Twain once remarked, "The difference between the right word and the almost-right word is the difference between lightning and the lightning bug." The same can be said of letters and memos that are geared closely to a particular situation, as compared with those that are not "quite right." The most obvious difference between a good letter writer and a poor one is the care with which the expert chooses words. Most people know with reasonable accuracy the meaning of the words they use. The difficulty arises because there are so many words in the language—and a lot of them are nearly synonymous. A first-draft writer will get the almost-right word much of the time. A careful person will search for a better word, one that more precisely expresses what the writer wants to say.

And so it is with this book. Finding the right letter or memo is not an automatic process. You will find it helpful to review briefly those models that deal with your primary purpose, particularly if several factors are involved. Suppose, for example, the letter you need to write should contain a thank you, a response to a request for technical data, and a clarification of corporate policy. Without this book, you would need to develop the complete letter, taking time to blend the elements referred to. However, reviewing the chapter headings that follow will lead you to letters that deal with all three topics. This search will take far less time than will creating the entire letter. In addition, each model offers alternate approaches so you can quickly adapt the phrases that most closely fit your needs. This is a unique feature, and one that substantially increases the book's utility value.

The table of contents lists the models in terms of the situations they deal with. A separate listing is also provided, organized according to whom the letter or memo is to be addressed. This will pinpoint those models that are geared primarily to five categories: (1) clients, (2) colleagues (outside your firm), (3) staff members, (4) contractors and suppliers, and (5) general.

Bear in mind that many entries are interchangeable—that is, a letter directed to a colleague in another firm may easily be modified so as to be suitable for use with a staff member.

Since they form the central theme of the book, we begin with the letters and memos themselves. If you wish, first, to review the fundamentals and a checklist of key factors in letter writing, turn to Chapter 10. Bear in mind, too, that many of your letters and memos will contain legal implications that should be carefully considered. Reminders of this appear at appropriate points; when in doubt, consult legal counsel.

G.E.P.

CONTENTS

LISTING OF MODEL LETTERS AND MEMOS RELATED TO KEY GROUPS

STAFF MEMBERS

Chapter 1

COMMUNICATING REQUESTS ON TECHNICAL MATTERS

- Requesting an Extension of Time   [1-1]
- Asking for Sustained Effort During Procedural Delays   [1-2]
- Requesting Earlier Completion of Job Assignment   [1-3]
- Requesting a Favor   [1-4]
- Asking for Business or Technical Advice   [1-5]
- Requesting Permission to Quote   [1-6]
- Requesting Colleague to Give Speech   [1-7]
- Asking for Ideas from Technical Staff   [1-8]
- Requesting Addition to Staff   [1-9]
- Asking for an Appointment   [1-10]
- Requesting Guidance on Staff Procedures   [1-11]
- Asking for Greater Staff Cooperation   [1-12]
- Requesting Answers to Listed Questions   [1-13]
- Requesting Acknowledgment of Letter or Memo   [1-14]
- Securing Reference Check on Prospective Staff Member   [1-15]

Although requests can be grammatically shaped in countless ways, the phrasing and overall tone can contribute to greater receptivity on the part of the reader, and eventually a more positive response. Difficulties inevitably arise when the writer is asking for something the recipient might view as unreasonable or impractical. In some instances, the request may be made in stages. A preliminary note to establish the basis for the request can be used, in a sense, as an overture to the main event. Since brevity in itself can be a persuasive factor, this is not usually the best way to go. Regardless of the circumstances, a good starting point is to anticipate the way you feel the reader will view the request, identify the factors that will help shape his or her response, and phrase your memo or letter accordingly. Another general rule would be to avoid a long-winded dissertation. After the stage is set, and a strong foundation for the request established, get to the point without further delay.

● *REQUESTING AN EXTENSION OF TIME*

General Guidelines

Be specific and concise when you deal with the request itself, and whenever possible, convey the thought you have also tried to look at the request from the reader's standpoint. Since many requests deal with the need for an extension of time in completing a particular project, Letter 1-1 provides an approach that can easily be modified to fit your situation. Note that the writer introduces his request with a restatement of the new system's value. In effect, he is saying that what he is offering is worth waiting for. He is also giving Mr. Johnson the basis for his explanation to others in his firm when informing them of the delay.

With a letter of this type, and others that go outside your firm, make certain you use the correct form of address. (See Chapter 11.)

Letter 1-1

OMEGA COMPUTER SYSTEMS, INC.
1212 Fairchild Avenue
Hillsdale, Illinois 00000

October 16, 19—

Mr. Alex Johnson, Systems Engineer
Worcester Power Supply Corporation
Wayside, Alabama 00000

Re: Disk Memory System Study

Dear Mr. Johnson:

The more we get into the Disk Memory System for use in your company, the more evidence we have that it will give you a much smoother operation, with far fewer breakdowns, thus saving your firm a substantial amount of money. I had hoped to get all the details into your hands by November 30, as we discussed, but we've uncovered a couple of additional design opportunities that should eventually make the system even more cost efficient.

To capitalize fully on these opportunities, it appears that January 3 would be a more realistic date for the presentation to your management, although you and I will be able to go over the complete setup by December 15. I hope you will not be inconvenienced by this delay, but I did want to let you know of this well ahead of time so you can plan accordingly. I'm sure you agree it is better to have every important detail covered, rather than have loose ends that could invite needless misunderstandings. Of course, if absolutely necessary, I could make a convincing presentation, even at this stage, but the entire system would not be shown as completely integrated. Nor would we be able to describe fully the additional improvements that now appear possible.

Please let me know your opinion as to which route we should take. I am hoping, of course, that the January 3 target date will be acceptable. Thank you for your consideration and understanding.

With best regards,

Joseph Smith, Design Engineer
Disk Memory Systems, Inc.

Alternate Approaches

(a)   Knowing that you appreciate quality of workmanship just as much as I do, I believe both of us will gain by your approval of the request contained in this letter. What we are aiming for here is a substantial contribution to your firm's profits, and several things have been uncovered recently that could help increase the size of that contribution.

(b)   You'll be glad to know that our efforts on the new Disk Memory System for your organization has uncovered several design opportunities that could increase the system's effectiveness. Although these opportunities require additional study before we integrate them with the total package, I believe the extra effort will most assuredly pay off later. This leads to the request that follows . . . .

(c)   You will recall that our preliminary conversations regarding the new Disk Memory System stressed the importance of a creative approach, one that would avoid the costly problems resulting from your present system. Surprisingly, since we're only in the beginning phase of development, new ideas have surfaced that appear to offer solid potential beyond our original expectations. We want to give these ideas the consideration they deserve. I suppose this type of situation is inevitable as we get further into our design of the new system, and it will be to everyone's benefit to capitalize on these opportunities to the fullest extent.

● *ASKING FOR SUSTAINED EFFORT DURING PROCEDURAL DELAYS*

General Guidelines

Letter 1-2 requests continuing action on a project, even though the decision promised earlier has not been made. It does not offer any assurance the eventual decision will be affirmative, but it does reflect the writer's desire to be forthright as well as his or her willingness to pursue the matter to an early conclusion. Delays of this type are not unusual and seldom require a long, infinitely detailed explanation. The reason simply needs to be stated concisely and, preferably, with some indication that the delay will ensure more careful consideration of key factors involved. Requesting a continuing work effort during a delay in the decision-making process should also reflect the writer's desire to resolve the situation as quickly as possible.

Letter 1-2

AJALAT ENGINEERING RESEARCH, INC.
1400 Baylor Avenue
Hampton, Virginia 00000

April 17, 19—

Mr. J. I. Kay
Diversified Instruments Co.
Westville, California 00000

Dear Mr. Kay:

I know I promised you a decision on the Hampton matter by the end of this week, but we've run into a number of troubling complexities here. They all relate to the patent application, and it's now obvious we need to do additional research on the legal aspects. This makes it imperative that we get qualified opinion

from Carl Jennings, our legal counselor, on how best to proceed.
I would follow through on this immediately if Carl were available, but he will not be back from Europe until next week. If it
is at all possible, would you be able to proceed along the lines
we discussed, with the understanding we will get back to you
within a short time after his return? We'd really appreciate it,
Jim. Please be assured our desire to move ahead on this is just
as great as it ever was. I'll look forward to your response.

Sincerely,

Richard Rowe
Chief Engineer

Alternate Approaches

(a)   The Fairfield matter appears to be much more complicated than we
originally imagined. This is particularly true of the information we must
provide in connection with the patent application. This key document
should (and will) get the careful consideration it deserves.

(b)   Our earlier discussions concerning the Fairfield situation were very
helpful, but getting down to the hard facts has uncovered a potential danger
that could cause us some real problems—specifically, the patent application itself. This fundamental information, if not developed properly, could
bring about a severe setback in our plans.

(c)   The progress we make on the Fairfield situation will be tied inextri-
cably to the patent application. It now seems we did not give sufficient
consideration to the legal aspects of this application, and it is essential that
we get expert legal counsel on this before going further. There's no point in
proceding, even on a preliminary basis, with design and production plans
until the application itself has been fully developed with the assistance, and
eventual approval, of our corporate attorney.

● *REQUESTING EARLIER COMPLETION OF JOB ASSIGNMENT*

General Guidelines

In comparison with the preceding situations, Letter 1-3 deals with the
opposite end of the scale, that is, a shortening of the time period originally

agreed on. As noted earlier, favors or requests are more likely to be granted if there is some benefit for the reader. The benefits need not be spelled out in endless detail. Just state the facts in a way that clearly implies the reader will ultimately gain by a positive response to the request.

Letter 1-3

ABRAHAM'S MANUFACTURING CORPORATION
112 Chestnut Street
Fort Davis, Maine 00000

May 3, 19—

Ms Doris Brown
Technical Illustrator
912 Broad Street
Peoria, Illinois 00000

Dear Ms Brown:

It isn't easy to ask a favor when I feel certain it will inconvenience you. This is compounded by the fact that we told you earlier you'd have two weeks to finish the artwork for our revised Training Manual. Unfortunately, because of recent developments I now find that we're not going to be able to allow you that much time.

The printer tells us he is so rushed that he must have all material, camera-ready, by May 20. Deducting the minimum of 6 days it will take our Production Department to assemble and edit the pages, that leaves just 11 days for you to finish the art and get it to us.

Can you do it in this time? And will you, as a favor to clients who appreciate your work?

On the other hand, if I'm pressing you too hard and you simply cannot finish the drawings in the shortened time, please tell me how much of it you can finish—and I suppose we'll have to make other arrangements for the balance of this work.

```
What do you say? Can you do the job for us on this shortened
schedule? It would really be appreciated. Please let me have your
answer in the next few days.

Sincerely,


Brad Jones
Production Manager
```

Alternate Approaches

(a)   Your preliminary sketches on the artwork looked just great. Now, because of an urgent message we have received from the printer, I must ask you to modify the original schedule. Completing the job in 11 days instead of the 14 we discussed earlier will be of enormous help to our production staff. If you can manage this, I feel certain Helen Walker (our production manager) will want to reciprocate this favor in the future.

(b)   I promised you earlier that I would talk with our production supervisor about future artwork assignments, and I have done so. She was most receptive, and I have asked her to get in touch with you when the next appropriate occasion arises. However, that is not the primary reason for this letter. We have run into a real problem with the printer's schedule, requiring completion of your present assignment in 11 days rather than 14.

(c)   The samples arrived, and ordinarily they would have been enough to make me ask for the finished artwork sooner than we agreed on earlier. You're doing an excellent job. But now we have an additional and much more urgent reason to make a slight reduction in the original schedule.

● *REQUESTING A FAVOR*

General Guidelines

There's much less need to be reluctant about asking for a favor if two factors are present: you are writing to a person who can fulfill the request without someone else's approval, and granting the request holds the promise of reciprocity in the future. Letter 1-4 embodies both factors.

After presenting the basis for the request, concentrate on the elements involved and avoid a plaintive or apologetic tone. Imply that you are aware favors are often discussed on a "two-way street" basis, and indicate your willingness to reciprocate at some point in the future.

Letter 1-4

WORTHINGTON ELECTRONICS CORPORATION
4612 St. Charles Road
Birmingham, Alabama 00000

April 20, 19—

Mr. Robert S. Smith
Southwest Industries
1211 Fifth Street
San Diego, California 00000

Dear Bob:

I wonder if I could ask a favor of you? A good friend and professional colleague, Ralph Jones of Modern Instruments, is going to be visiting San Diego for two weeks beginning May 12. He is a complete stranger to your city and will be by himself. When he mentioned this, I impulsively told him I knew somebody there. Would it be all right with you if I have him call you? And could you spare the time to show him around Southwest Industries?

I think you would enjoy Ralph's company. He's a pleasant, capable guy and an exceptionally sharp engineer. His specialty is quality control at Modern Instruments. Anything you can do to make his stay more enjoyable would be greatly appreciated by me. And, of course, I'd be delighted to return the favor any time you might want me to.

I wish I were coming along too, but I'm swamped with work because this is our busiest season. Please let me have your answer within the next week or so.

With best regards,

John

Alternate Approaches

(a) I was reminded recently of our past conversations on quality control. The reminder came in the form of a good friend and professional colleague, Ralph Jones. Ralph is with Modern Instruments in Birmingham and directs the firm's quality control operation. Coincidentally, he told me at lunch yesterday of a forthcoming business trip he'll be making to San Diego. This will be his first trip to your fair city, and since he'll be a complete stranger to the area, I suggested he give you a call when he arrives on May 12.

(b) Although we've never discussed quality control per se, it seems that many of the things we've talked about have a close relationship to the subject. This thought came to mind yesterday when I was having lunch with Ralph Jones, quality control director for Modern Instruments here in Birmingham. Ralph is a long-time friend, and I have great respect for his technical expertise. Hearing about some of the innovations he introduced to Modern Instruments made me feel you and Ralph have a number of things in common. As a result, I was pleasantly surprised to learn he's planning a business trip to San Diego (his first time there) and expects to stay two weeks, arriving on May 12. Upon hearing this, I asked him to give you a call.

(c) Asking for favors is something I try to avoid. In this case, however, I'll make an exception because I believe the result will be beneficial to all concerned. Ralph Jones is a good friend and professional colleague with superb training and experience in the field of quality control. As a key executive with Modern Instruments he does a considerable amount of business traveling and told me yesterday he'll be going to San Diego (for the first time) on May 12. When I mentioned you, he said he'd be most grateful if you could manage to show him around the Southwest Industries operation.

● *ASKING FOR BUSINESS OR TECHNICAL ADVICE*

General Guidelines

Making a direct request for business or technical advice presents its own special set of problems, particularly when a fee may be involved and the amount is not known at the outset. Letter 1-5 is designed to pave the way for a more comprehensive discussion, one that would lead to appropriate and specific arrangements concerning possible fee, time period, dates, and so on. Your reasoning, as expressed in the letter, should also explain why you feel the recipient is especially well qualified to advise you on the particular problem at hand.

Letter 1-5

ELECTRONICS COMPONENTS, INC.
406 James Boulevard
St. Louis, Missouri 00000

April 7, 19—

Mr. Jacob Green, President
Engineering Consultants, Inc.
910 Broadway
Chicago, Illinois 00000

Dear Jake:

As you know, we have a small electronics part supply store in this city doing a volume of $3 million, which is only fair for the size of the store and the location. After listening to your talk last month at our regional meeting, we've made a real effort to broaden our horizons.

We did take on one additional parts supply shop about six months ago. It is doing fairly well, but I've had difficulty locating another. Recently, we learned there is one available, but I have some doubts about it—chiefly because of the location. It is fairly well situated with plenty of parking space, but we're concerned about the distance between the location and our main parts supply warehouse.

My staff has, in a sense, contributed to the dilemma because they are about evenly divided as to whether we should go ahead on this. I'm wondering if I could ask a favor of you. Could you possibly get down for a few days, look over the situation, and give me your opinion?

We'd put you up at the nicest hotel in the area, and it would also be a good opportunity for us to get together socially. I'm

sure my wife would enjoy meeting Mrs. Green if she were able to
join you on the trip.

Let me add that I would certainly not hold you responsible for
any decision we might finally make on this. I just feel that I
need the advice of an experienced professional in this field. May
Janice and I roll out the red carpet?

With best regards,

Harry

Alternate Approaches

(a)   Although it has been over a month since I listened to your talk at
our regional meeting, I can't get some of the things you said out of my
mind. Nor do I want to. It was one of the best presentations I've ever heard
on business expansion methods. One tangible result has been the new
plans we're developing for additional electronics parts supply stores. Quite
frankly, we're excited about the potential these plans offer, but I don't
want to go too far or too fast without the advice of an acknowledged expert
in this field.

(b)   I suspect you received a number of compliments regarding your
presentation on tested ways to ensure business growth. If you didn't, you
should have. It sparked a number of ideas we're exploring in our own
business, and even though we're just in the preliminary stages, I think some
of your ideas would fit in beautifully with our plans. Problem is, I doubt if
anyone is as qualified as you are to judge whether we're on the right track.
Since the quality of long-range planning depends so much on the action that
precedes it, we're especially anxious at this point to have the kind of guidance you could provide.

(c)   Your remarks on business expansion in technical product areas had
a profound effect on members of my staff and myself, so much so, that ever
since your presentation we've concentrated on developing growth plans
based on the "10 Rules" you referred to. Our present electronics supply
store does a volume of $3 million, and in my opinion, that's only fair
considering our location and the size of this city. If we're ever going to add
additional retail outlets, and we are determined to do just that, the plans
presently being developed must be reviewed by someone whose judgment
we would welcome—and respect. Based on the superb presentation you

made at our regional meeting, I don't think we could find anyone better qualified to fill this role than you.

● *REQUESTING PERMISSION TO QUOTE*

General Guidelines

Letter 1-6 represents the type of letter that is mandatory if your paper or manuscript soon to be published contains direct quotations, paragraphs, condensations, synopses, charts, tables, or illustrations from copyrighted publications. You will need to secure written permission from the appropriate individual who represents the copyright holder. This requirement may also apply to unpublished papers, transcribed speeches, or dissertations you draw from. Send the written request in duplicate, so that the addressee can keep a copy and return the original (signed) copy to you.

Be specific about the material you wish to use, the publication in which it appeared, and the page number, if appropriate.

Indicate exactly how, and why, you would like to use the material.

Letter 1-6

BENSON HIGH-TECH PRODUCTS CORPORATION
412 Canyon Avenue
Lester, Illinois 00000

April 7, 19—

Ms Helen Gabacia, Editor
Technician's Journal
800 Century Blvd.
New York, New York 00000

Dear Ms Gabacia:

We are preparing an article on high-tech investment opportunities to be published next January for distribution to our membership. Because the interesting article in the September issue of

your journal is relevant to this topic, we would like to reproduce it in our forthcoming newsletter. The title of the article
is "New Developments in ICs," written by Donald Smythe. The author, as well as your fine publication, would receive full credit
in the issue that uses the article.

We would appreciate your permission to use this material. For
your convenience, a release form is enclosed. Please sign and
mail the original in our self-addressed stamped envelope, re-
taining the copy for your files. I'm sure you're aware of the
pressure that deadlines can exert on publication schedules, so
we'd greatly appreciate your response within the next week or so.

Thanks very much for your cooperation, Ms Gabacia.

Sincerely,

John Kinella
Research Staff

Alternate Approaches

(a)   High-tech investment opportunities will be the subject of a forthcoming meeting to be held by our local chapter of the Engineering Management Association. The article by Donald Smythe beginning on page 42 of
the September issue of your journal was both well written and timely. I do
hope you will permit us to reprint this article, with full credit of course, in
next month's issue of our local newsletter (sample enclosed).

(b)   As editor of the newsletter distributed to members of the local
chapter of the Engineering Management Association, I'd like to express
appreciation for Donald Smythe's interesting review of new developments
in integrated circuits. This appeared in the September issue of your journal,
beginning on page 42. The subject ties in nicely with an article we're
preparing on high-tech investment opportunities, to appear in the January
issue of our newsletter. May we have permission to reproduce Mr. Smythe's
review in this article?

(c)   Your journal has contributed many useful ideas and concepts to our
organization. This was most recently illustrated by Donald Smythe's article
on New Developments in ICs, beginning on page 42 of the September issue.
It ties in with a presentation we plan to make to members of our Management Association on high-tech investment opportunities. In addition to

commending the journal for the high caliber of its content over the years, we'd like to reprint Mr. Smythe's article in our next newsletter. Your permission would be appreciated. We will, of course, be glad to use the credit line you suggest.

● *REQUESTING COLLEAGUE TO GIVE SPEECH*

General Guidelines

Letter 1-7 capitalizes on the fact many people are pleased when they are invited to speak at an association meeting or seminar. A gracious letter giving the necessary details is often all you need to persuade the recipient to accept your invitation. Give the prospective speaker a precise description of the subject matter or particular topic you would like him or her to cover.

Briefly review the circumstances leading up to the request, and compliment the individual on his or her qualifications to talk on the subject.

State the time, place, and desired length of the speech. If the question is likely to arise, also indicate whether the chapter, club, or association provides any fee or honorarium.

Emphasize your membership's interest in the topic the speaker would cover, and convey the assurance he or she would receive a warm welcome.

Letter 1-7

R & W ENGINEERING RESEARCH
445 Carriage Hill Road
Tarryville, Vermont 00000

August 8, 19—

Mr. Elliot Schwartz
845 Birchtree Road
Syracuse, New York 00000

Dear Mr. Schwartz:

The subject of "effective technical communications" is of interest to all professionals. I suspect this is particularly true of those in every branch of engineering. You, of course, are obviously aware of this because of the book and several articles you've written on the subject.

As program chairman of the local Professional Engineers Association, I would like to invite you to be the featured speaker at our November 17 meeting. It would be our pleasure to have a person of your caliber and experience give us a presentation on this important topic. It would also be timely, because we recently conducted research in this area that I'd like to share with you. The honorarium we can offer is moderate, but it will cover your traveling expenses, and we'll also be glad to distribute circulars describing your book during the week prior to November 17.

Our dinner meetings are held in the Fireside Room of the Hotel Cleveland. The cocktail hour starts at 6:30 P.M., dinner at 7:30 P.M., and the presentations usually begin around 8:15 or 8:30 P.M. The speakers generally make presentations that range from 30 to 40 minutes, followed by a question and answer period. We seldom have fewer than 125 members in attendance, and on this occasion I feel reasonably certain there will be a larger number.

Provided you accept our invitation for November 17, I believe you'll receive a warm and most enthusiastic welcome from members of our organization. In any event, could you let me have your response within the next week or two?

Thanks very much, Mr. Schwartz.

Sincerely,

Alternate Approaches

(a)   Your article in this month's *Data Network* magazine was exceptionally interesting and informative.

Concurrently, a number of people have told me about the outstanding talk you gave to the regional meeting of the Programmer's Association. So many of my colleagues have mentioned this, I would be delighted if you could speak to our organization at next month's meeting. Our program committee is hard at work on arrangements for this important event sponsored by the Professional Women in Data Processing. We're inviting several outstanding computer professionals to attend.

(b)   A talk based on your published article would make a most interesting presentation on the growing computer movement abroad. I am, therefore, extending a cordial invitation for you to speak informally to the next meeting of the American Computer Association.

This will be held on November 18, at 8:00 P.M. at the Biltmore Hotel on Beeker Street in Tarryville. Your experience with international computer applications is of particular interest to our membership, and I know they would be most attentive to your remarks on the subject.

(c)   I was fortunate enough to be in the audience on Tuesday, March 11 when you spoke to the Atlanta Engineering Association. I have never heard the emerging opportunities in fiber optics spelled out so clearly. It was a most impressive talk. So much so, in fact, that I hope your schedule will permit you to speak on this subject at our annual meeting on November 14. Provided you are willing to do so, I'd like to issue a special bulletin to our members. Could you let me have your answer in the next few days?

● *ASKING FOR IDEAS FROM THE TECHNICAL STAFF*

General Guidelines

Memo 1-8 deals with a topic that represents a challenge to most supervisors: securing constructive ideas and useful feedback from staff members. The need to do so is particularly important with matters involving technical procedures and/or projects. Employees can be an invaluable source of new ideas when they are effectively motivated to offer them, but an occasional request or the traditional "suggestion box" approach seldom produces consistently satisfactory results. A memo or letter, if sent at regular intervals with varying approaches, can provide the regular reminders needed. It will also serve to assure staff members of your continuing interest in their suggestions and ideas on new procedures, products, techniques, and the like.

Memo 1-8

April 18, 19—

TO: Engineering Supervisors

FR: Robert Olson, Manager

Some of you have talked to me about ways to improve staff performance. There is no easy way to change corporations or people

unless they want to change. An engineering staff is comprised of
people. It takes its personality from what they do for, with, or
against each other.

There is no question that improvements in scheduling and work
procedures are possible. The easy thing would be to ask each of
you to make a detailed presentation of your thoughts on the matter. Instead, I'd like you to establish priorities on the suggestions you have. Then, send me a one-page memo that deals only
with your perception of the one most needed improvement. If this
proves productive, you can be sure we'll get to any others on
your list.

Please let me have your memo on May 14, which is the day preceding our next staff meeting.

Alternate Approaches

(a)   We've had several talks recently about the important objectives we
need to accomplish by the end of this year. Several ideas resulted that could
help accelerate our progress, and now I think we've reached the point where
our best thinking needs to be pinpointed and put down on paper. If we can
develop a sharp focus on those things that not only need to be done but *can*
be done, this could add up to a productive step forward. Trouble is, we've
kicked around so many ideas that some of the best ones may have been lost
in the shuffle. Which leads me to the purpose of this memo.

(b)   At our last supervisors' meeting I began to wonder why those in
attendance were trying to solve problems that could more effectively be
resolved by staff members. This thought was immediately followed by the
realization we had little or no input from staff members simply because we
had not asked for it. As you know, this led to the meeting I had with you last
week. A lot of ideas were discussed at that meeting mainly dealing with the
three new products we're developing. It was a good meeting, primarily
because it laid the groundwork for this request. Specifically, we've had time
enough to consider the various approaches discussed. Now, let's distill our
individual thinking so we can set priorities on those things that need to be
done first.

(c)   One of our colleagues in the Quality Control Department recently
remarked that "New product ideas are a dime a dozen, and most of them
are not worth that much." I'm glad he will not be attending our staff
meeting on new product development to be held at 3:00 P.M. on Friday of
next week. At this meeting I want us to have a realistic but free-wheeling

discussion. All ideas will be considered. Come prepared to offer your suggestions in a relaxed, informal atmosphere, and don't hesitate to come up with "far-out" ideas as long as you think they offer solid potential. Following this discussion I'll ask you for a one-page memo dealing only with what you consider to be the one most potentially profitable idea advanced at this meeting. We'll get to the other ideas later on, but initially let's reduce things to a "first-things-first" basis.

● REQUESTING ADDITIONS TO STAFF

General Guidelines

Memo 1-9 provides a tested approach to the dilemmas caused by insufficient personnel. Asking for additional workers is seldom easy. One reason is because of the action's inevitable effect on budgetary matters, among other things. This increases the need to establish a factual, convincing foundation for the request. The situation also represents one of those occasions when a two-step approach might be considered. For example, describe the specific problems (and costs) resulting from insufficient personnel, and indicate the current steps being taken to avoid the need for additional staff. Then, if the steps do not lead to a solution of the problem, the groundwork has been laid for your request. It is also important to be specific about the ways in which additional staff will contribute to increased profits. Solving production problems, for example, is one thing, but translating this into the ways it relates to the profit picture will add a more compelling note to your request.

Memo 1-9

April 9, 19—

TO: John White
      General Manager

FR: James Bradford
      Director, Systems Engineering

John, as you know, the overall volume of our business has increased 50 percent within the last few months. Nobody is

complaining about the increase, but I had a meeting yesterday
with our systems people, and they tell me the surge in business
has created serious problems in the Shipping Division. Every bit
of slack has been taken up in this department, and we've instituted time-saving methods wherever possible. Despite this, we're
falling farther and farther behind on delivery schedules.

All the facts I can gather convince me we've reached the point
where it's imperative we add at least two people to the shipping
crew. Delays on outgoing orders, if they get much worse, will undoubtedly hurt us with customers and undo many of the benefits
resulting from the increased volume of business. Compounding the
present situation, our present staff is becoming disgruntled
with the constant overload, and it's possible we'll lose some of
our best people.

The attached cost breakdown and related information will give
you a clear picture of the effect this will have on our overall
budget and expense projections. In view of these figures, and the
urgency of the situation, I'm requesting permission to hire two
new shipping crew members now, with the understanding that if
this is insufficient, approval will be given to hire a third by
the end of May. Additional documentation will be provided in the
event the third person is needed. I know this action goes beyond
our present budget, but the circumstances appear to me to provide
adequate justification. I'm convinced this additional manpower
would be good for the business both now and in the long run. May
I please hear from you as soon as possible?

Jim

Alternate Approaches

(a)   As you know, a surge in the sale of our Benton product line occurred
six months ago, and it has grown consistently ever since. We are now 44
percent ahead in sales volume as compared with the same period last year.
To compound this (from a manpower standpoint), we were forced to transfer one of our employees to the Quality Control Department last month. So,
we're actually in a position of trying to cope with a substantial increase in
production requirements, with fewer staff members. I have attached a list
of the procedures we've implemented in an attempt to avoid requisitioning
two new employees. Despite this concerted effort, we're losing ground

rapidly. It now appears inevitable that delays in production will affect our profit picture if we don't add to staff in the near future.

(b)   The situation facing our production department is becoming more urgent. Our level of manpower in this area has remained constant over the past six months, while output has increased by 32 percent. Frankly, I believe we not only owe this department a round of applause, we owe them our support in the form of at least one new staff member (see job description attached). We need to do this before the situation becomes critical. I have attached statistics that pinpoint the problem areas. Our Customer Relations Department provided me with an assortment of letters regarding delayed shipments, and these are also attached. Note there is not one letter regarding product quality. They deal with genuinely unreasonable delays. There is no way we can avoid a negative effect on profits in this area if we don't take appropriate action quickly. And I believe appropriate action involves adding a new person in the near future.

(c)   I have attached the production quotas assigned to each member of our staff in January of this year. After you review this, please note our current figures regarding productivity. As you will see, eight months later each member of the department is performing well above the quotas set earlier. This has not been managed without considerable cost, as indicated by the overtime figures. We are kidding ourselves if we think this situation can be allowed to continue without its having a harmful effect on staff members, and profits. Consider also the comparison among overtime costs, customer complaints, delayed shipments, and the cost of the new employee I'm recommending. In essence, we're draining our resources instead of building greater strength into this department. Taking the latter course would help ensure an increased contribution to profits in the future.

● *ASKING FOR AN APPOINTMENT*

General Guidelines

Letter 1-10 will help pave the way for constructive meetings and staff discussions. Letters and memos of this type generally fall into one of two groups: asking for an appointment or asking someone to visit you at your location. State the purpose of the meeting, but avoid going into extensive details unless it is essential that you do so. When meeting with a staff member on several technical matters, it would be well to spell out the agenda to be discussed. Be specific regarding time, place, and the date of the appointment. Or, if appropriate, ask the person to whom you are writing to suggest these details. When there is any question regarding the person's response, ask for confirmation of the appointment.

Letter 1-10

CENTURY ELECTRONIC DEVICES
Century Park
Brooks, Massachusetts 00000

April 9, 19—

Ms Yvonne Kern
19 Branch Avenue
Lester, Massachusetts 00000

Dear Ms Kern:

Having just returned from Chicago where I conferred with Robert Greenwood, chief mechanical engineer for Electromatic Corporation, I believe it would be in both our best interests to discuss the various topics Robert and I talked about. It appears there are opportunities here that could contribute to engineering designs you are presently involved with, and Bob Greenwood would also be a potential customer for some of the products eventually resulting from these designs.

Please call early next week and let me know when and where it would be convenient for you to join me in discussing this.

Sincerely,

William Fulton

Alternate Approaches

(a) I will be in New York on August 1. Would it be possible for me to meet with you at 11:00 A.M. to discuss the design specifications for the X100 receiver?

(b)   Would you please join me in my office on September 11, at 9:30 A.M. I would like to discuss your present work assignments and the possibility of a temporary reassignment of some of your duties to the Data Processing Department.

(c)   Your design specs for the X111 project are well drawn and I believe they bring us close to the decision-making stage. We can discuss the few questions I have if you will see me on Tuesday, March 12, at 10:00 A.M. in my office.

(d)   We need to have a detailed discussion of your plans for the RM12 series. I am primarily interested in three things, although I'd like an overview from you on the total program. The three topics I'd like you to cover in detail are current production problems as reflected by your March monthly report, your specific proposals for correcting these problems, and your projection of production output to be achieved on or before September 1.

(e)   I have several profit-related ideas I'd like to discuss with you. My guess is that it would take no more than a half-hour of your time, and I would be glad to visit your office at your convenience. Although I will be in New York on August 1, I will arrange another date if this is not a good time for you. Please let me know what date would be best. The ideas I have relate principally to design flaws in our M12 transceiver.

● *REQUESTING GUIDANCE ON STAFF PROCEDURES*

General Guidelines

Memo 1-11 will help minimize the reluctance many people have in requesting guidance or assistance, particularly if it relates to matters under their supervision. Depending on the phrasing, however, the recipient may feel complimented that he or she is considered capable of contributing to the solution of a particular problem. When requesting cooperation in resolving a particularly difficult situation, indicate clearly and concisely why the request is being made. Let the reader know why you came to him or her for help, and how that help will be used. Whenever possible, stress the mutual benefits resulting from a successful resolution of the difficulty described in your memo or letter.

Memo 1-11

April 9, 19—

TO: Allen Grant

FR: Mary Peterson

Allen,

You are aware of the reduction in staff resulting from current
economic conditions. As a result, we're finding it impossible to
complete all the monthly reports required by the headquarters
office on time. Several steps have been taken to consolidate related information, to redistribute work loads, and to eliminate
any duplication of efforts. Our staff members fully realize the
importance of making every minute count, particularly in the
current circumstances, and we're continuing to search for additional ways that will help expedite our reports.

In spite of all these efforts, we're continually falling behind.
This forces me to ask a favor. I really believe you can help.
Would you please visit our office to review the various reports
with us? It seems likely there are some that might be eliminated
or at least reduced in complexity. It's also possible there are
others that could be combined. Then, perhaps you could persuade
the headquarters office that these reductions in the reporting
work load would not eliminate any essential information they
need.

We would appreciate a visit from you, Allen, and believe the company as a whole would benefit from this. Please let me hear from
you soon.

Mary

Alternate Approaches

(a)   Although we have reduced our staff by three people, there has been
no reduction in the number and complexity of reports we must complete
monthly. Previously, even though we had a full staff, this took an inordinate

amount of time. The situation was compounded by the fact we never seemed to know whether *all* the reports were serving a valid purpose. Now, with the current shortage of manpower, we're being snowed under by the facts, figures, and other data that must be compiled on a daily basis. Productivity is being seriously affected, and, quite frankly, we need the kind of opinion and help I believe you could provide.

(b) Staff members in this department have recently reinforced my concern regarding the number of daily, weekly, and monthly reports we're required to submit. The recent drop in production provides further evidence that this has become a serious problem. Please review the production figures attached, covering the last four months. Note that in January we were up to full staff, but in early February we lost two experienced engineers and one lab technician. There has not been, however, any reduction in the reports that must be filed with your department. I would be less than candid if I didn't tell you we need your help on this, because I'm convinced that cutting down or combining some of these reports would help both our departments save time and money.

(c) To offset the recent reduction in our staff we have redistributed work loads and taken other steps to eliminate duplication of effort. There is one area, however, where I have been reluctant to take any action arbitrarily, primarily because it relates to the extraordinary number of reports we must fill out and send to corporate headquarters on a regular basis. Corrective action, if it is going to be effective, must come from both our departments. We really need to get together to see if some of these reports can be combined or eliminated altogether. Your help is needed to avoid any unilateral action that would not be in your, or our, best interests.

● *ASKING FOR GREATER STAFF COOPERATION*

General Guidelines

Memo 1-12 provides a mechanism designed to encourage a greater sense of teamwork among staff members. A wide range of factors and events can affect the degree of cooperation among employees. A new supervisor, an offensive co-worker, excessive or inadequate supervision, corporate takeovers, and work reassignments are just a few of the things that affect morale and contribute to dissension. While departmental meetings and individual conversations with staff members are desirable, any request of this type to the entire staff should ultimately be put in writing. There are two advantages: everyone receives the same message at the same time, and making it part of the written record can give it more lasting impact. It is seldom sufficient to make this type of request a "one-shot" affair. Follow-ups are often necessary, either in the form of a stronger, more insistent tone or an expression of appreciation for improvements that have been brought about.

Memo 1-12

April 9, 19—

TO: Staff Supervisors

FR: Robert Haywood, Chief Engineer

RE: The Recent Work Stoppage

This is a difficult memo to write, even though I realize that much of what I want to say is already understood and accepted by many of you on an individual basis. Naturally, I'm delighted that we are all back on the job again. We have been through a difficult period—so difficult, in fact, that I suppose the word "together" can apply only in the physical sense at this time.

Nevertheless, the nature of our work demands unity. Unity contributes to cooperation, teamwork, and profits. Whatever feelings we may have about individuals and issues, our job here is to help design quality products in the most efficient way, and that job requires a team effort. The very nature of our assignment makes it essential that we work together with mutual respect, despite any personal differences.

Little is to be gained from an atmosphere hostile to productivity or personal growth. Prior to recent events, we enjoyed a reputation for working together smoothly and efficiently. The days and weeks ahead can enhance that reputation. We need to get on with the job of assuming our individual responsibilities in order to capitalize fully on the opportunities available to all of us.

One month from the date of this memo, I will expect a brief, written report from each supervisor summarizing progress being

made on our projected schedule for the X-400 project. This report
should follow the same format as is used in the quarterly status
reports.

Thank you.

Alternate Approaches

(a)   Our recent experience with the X-100 project taught many of us a
lesson. Actually, several lessons. But there is one in particular I'd like to
share with every member of our staff: the need for greater teamwork. No
one appreciates the fact more than I do that we are a department composed
of individuals. I respect each member of our staff for his or her individual
abilities and desire to succeed. What I am asking for (and will insist on) is a
greater effort to combine and coordinate our individual skills in order to
ensure a smoothly functioning department as a whole.

(b)   On three occasions in the past week we failed to meet our departmental quota for a reason I find hard to believe. Production dropped in
each case because the employees involved simply did not communicate
essential data to each other. If the lapses were intentional, it would, in one
way, simplify corrective action. But investigation convinces me it was an
oversight. An expensive oversight, however, and one that we should be
determined to avoid in the future. All this, as nearly as I can gather, seems
to boil down to an absence of teamwork. This should be a reminder to all of
us that working together to meet our departmental goal is a vital obligation
on the part of each staff member. Teamwork never happens accidentally.
It's necessary to work at it. Working together, smoothly and effectively,
should be a top priority for each member of our staff.

(c)   Every person on our staff is a team member. Several incidents have
occurred recently that cause me to wonder if we need to be reminded of
this more often. Our research on the X-100 project would have been
simplified and substantially accelerated if there had been greater evidence
of teamwork. Quality Control would not have rejected over 40 units if they
had known we deliberately substituted different capacitors because of
design changes. This was a good and necessary improvement in design
specs, but no one told the Quality Control people. There have been other
instances of this type recently, when we did not communicate or cooperate effectively with each other, and with other departments. I would like
the supervisors to meet with me at 10:00 A.M. next Monday to discuss
this subject.

● *REQUESTING ANSWERS TO LISTED QUESTIONS*

General Guidelines

   Memo 1-13 illustrates the fact that when asking several precise questions in one letter or memo, you can ensure an equally specific response by numbering the questions. The completeness and practical value of the reply you receive depends largely on how persuasive you are in expressing your need for the information, and how precisely you phrase your questions. Make your inquiries as clear and direct as possible. Give assurance you would be glad to reciprocate in the future. Express your appreciation for the reader's response, and do not "load" each question in a way that would require a long, multifaceted answer. If the reader is asked to supply only brief, factual responses, he or she will be more likely to provide the information you need.

Memo 1-13

March 2, 19—

TO: Eloise Carpenter

FR: John Desmond

I know that favors can go in two directions, and if you can help us with this request, I'd be glad to reciprocate at some point in the future. Specifically, we're considering the installation of a small computer to replace hand posting of data compiled by our engineering staff. I understand you have had experience with the Conrad 103 system. Could you help us by answering the following questions? A brief answer is all we need.

1. How long have you had the system and how frequently has it required adjustments or repairs?

2. Has the use of this system reduced expenses? If so, in what ways and to what extent?

3. Do reports and other data get to your managers at an earlier calendar date than before the system was installed? In other

```
words, please describe any time savings that resulted from
this purchase.
```

I would really appreciate your response to these questions.

John

Alternate Approaches

(a) It appears we're about to follow your lead by investing in the same system you purchased a few months ago. The Conrad 103 seems to offer the features we need to automate data that are presently hand posted by our engineering staff. Frankly, we're not as sure as I'd like to be that this is the ideal system for us, but I won't impose on you by asking you to contribute to a decision that only we can make. There are, however, a couple of questions you would be much better qualified to comment on than we are. They have to do with your own experience in using this computer. No need to spend a great deal of time on these few questions. Your brief responses will be just fine. I hope there'll be some way I'll be able to reciprocate in the future, and thanks so much for your help.

(b) For some time now we've been considering the purchase of a Conrad 103 system. I won't take up your time by going into detail on the ways in which we'll be using the system, because the purpose of this letter is to focus on *your* experience with this computer. Conrad's sales representative told me about your purchase of the unit, and because I believe our organizations have much in common in terms of dealing with technical products, I would value your responses to the few questions listed in this letter. Your brief, candid comments will really be most helpful, and I would be glad to reciprocate whenever there's an opportunity for us to do so.

(c) A mutual friend, Alan Groves, recently mentioned that you had purchased a Conrad 103. I was glad that he told me this, because for some time now I've been searching for someone in our profession who has had experience with this machine. Our own interest relates to engineering data that are currently being hand posted, and it appears the Conrad system could not only handle this a lot faster but could organize the data and feed it back to us in various helpful ways. Naturally, we've concentrated first on whether it's the best computer for our purposes. But there are always questions concerning reliability, maintenance costs, and the experience of others in the use of this system. With the latter in mind, I would be especially grateful for your responses to the brief questions that follow. You may be sure I would welcome an opportunity to reciprocate whenever we can be of help to you. Just a frank, brief answer is all we need, and I've enclosed a stamped self-addressed envelope to simplify matters.

● *REQUESTING ACKNOWLEDGMENT OF LETTER OR MEMO*

General Guidelines

Letter 1-14 provides a brief, diplomatic way to draw a response from the individual who has not replied to an earlier inquiry. There is a reluctance on the part of some people to send a follow-up note when someone has not responded to an earlier communication within a reasonable length of time. Whether the length of time involved is "reasonable" or not depends on the urgency of the situation, and whether the original communication made it clear that a reply was necessary. In general, follow-up "notes" are just that: brief inquiries that courteously ask for a response to an earlier communication. It does little good, at least initially, to use an abrupt or contentious tone. This could prove embarrassing if, for example, the original letter went astray and was not delivered to the appropriate person. Summarize the nature of the original communication, and do so in a way that stresses its importance while underscoring your need for the kind of response it requested.

Letter 1-14

FRAILY CORPORATION
1260 Madison Avenue
Hampton, Virginia 00000

July 9, 19—

Mr. Harvey Walker
Chemical Industries
Belletown, New Jersey 00000

Re: Kenyon Contract

Dear Harvey:

I wonder if there has been some kind of slip-up in the mail service. I wrote you a query two weeks ago concerning the Kenyon

contract we're considering, and we've had no response. This is
becoming a matter of some concern, because we're getting close
to the time when a firm decision must be made. In the event you
did not receive my detailed letter, I'm enclosing another copy.
As soon as you have read and applied your best thinking to this,
please give me a call at the Harrisburg office. I'll be there
until August 5. On August 6 I leave for Toronto.

With best regards,

John Maxwell
Production Planning

Alternate Approaches

(a)   The fact that I have not heard from you in response to my letter on
the Kenyon contract (sent two weeks ago) puzzles me. It's unlike you to
delay action on such an important matter, and I'm now wondering if you
received my letter of March 22. To play it safe I've enclosed another copy
and would appreciate it if you'd go over this just as quickly as possible and
get back to me with your comments.

(b)   We've come to a standstill on the Kenyon matter because there's no
way we can move ahead on this until we have your detailed reply to my
letter of March 22. The situation is becoming urgent, Harvey, and I would
appreciate your attention to this. In the event the earlier letter went astray,
the enclosed copy will acquaint you with contract clauses that need to be
changed. As soon as you've gone over this (preferably Thursday or, at the
latest, Friday), please give me a call.

(c)   The last time we talked on the telephone, you assured me my letter
of March 22 on the Kenyon contract would get top priority. It has now
been two weeks, and we're still waiting for your response along with the
circuit boards needed to fulfill the contract terms. When I called you this
morning, your secretary was not certain as to when you'd be back in the
office. I really don't want to badger you about this, Harvey, because
you've always been prompt and cooperative on those occasions (fortunately, quite rare) when we've needed help urgently. A copy of the March
22 letter is enclosed. I will need your response in writing, so after reviewing this please send me your written recommendations via registered mail
no later than April 4.

- *SECURING REFERENCE CHECK ON PROSPECTIVE STAFF MEMBER*

General Guidelines

Although a request similar to Letter 1-15 follows a relatively standard format, the responses can cover a broad range. Some firms have a policy of not releasing information on former employees, others limit their comments to a few words, for example, "performance satisfactory." While, as indicated by the "Alternate Approaches," a brief questionnaire may be sent, this may be unproductive if the questions are not, indeed, briefly stated. Other firms will not accept reference checks via telephone, and for good reason. The sample letter takes a "nondirective" approach by asking the recipient simply to summarize his or her opinion of the former employee's performance, one advantage being that it does not limit the individual's response to questions posed by the writer. Whatever approach you take, make it as direct and concise as possible.

Letter 1-15

AMES CORPORATION
12 Benton Street
Carswell, Missouri 00000

March 12, 19—

Mr. Harold Kern, Chief Engineer
Baylor Industrial Corporation
444 Hansen Boulevard
Willoughby, Michigan 00000

Dear Mr. Kern:

One of your former employees, Mr. Robert Wells, has applied for a position on our technical staff.

According to Mr. Wells's application, he was employed by your firm from March 1986 to January 1989. He also listed your name as

a personal reference. I would very much appreciate it if you
would summarize your evaluation of Mr. Wells's job performance
while he was with your firm. Your response will be held in strict
confidence, and we would be pleased to reciprocate if such an occasion arises in the future.

Thank you very much, Mr. Kern.

Sincerely,

Donald A. Wesley
Technical Staff Manager

Alternate Approaches

(a)   We are seriously considering the employment of Robert Wells for a
position on our technical staff. Mr. Wells worked for your organization as
quality control supervisor for three years and listed your name as a personal reference. The brief questionnaire enclosed will take but a few moments, and your comments will be most helpful in our evaluation of Mr.
Wells's qualifications. You may be assured your comments will be held in
strict confidence.

(b)   We recently had an application for employment from one of your
former staff members, who spoke very highly of your organization. I refer
to Mr. Robert Wells, who, I understand, left your firm when he relocated to
this area. I would appreciate your comments regarding such matters as his
overall job performance while with you, attendance record, compatibility
with other staff members, and any additional information you feel would
be helpful. Thank you, Mr. Kern.

(c)   Since we both occupy positions in a highly technical field, I suspect
we share a mutual concern regarding the employment of new staff members. An applicant's previous experience can be a crucial factor in the
employment process, and that is the reason I am writing to you. Mr. Robert
Wells, who was on your staff for three years, recently applied for a position
with us. He assures me he had a satisfactory work record with your firm,
and expressed the belief you would confirm this. Would you please send me
your comments on his overall performance while with you? Your remarks
will be treated confidentially, and I look forward to hearing from you.
Thanks very much, Mr. Kern.

A WORD ON STYLE

Edit Your Own Writing

Editing is hard. You know what you intended to say, but the poor, bewildered reader doesn't. The reader has only one way to get your message, and that is from the words you put on paper.

A good writer is sometimes a good editor, but not always. The skills are essentially different. The best way to self-edit is usually not possible for a busy letter writer. That is because the best way is to wait awhile, to put your writing aside and forget about it. Then when you come back to it, you can see it anew, and you can edit it; you can cut the nonsense and the fuzziness. But a letter writer can seldom wait.

If your secretary is an editor, you are in luck. If not, you will have to cultivate an ability to reread critically, impersonally, instantly. It can be done, but it isn't easy.

Chapter 2

PROVIDING, OR SECURING, TECHNICAL COOPERATION

The purchase of equipment and contracting for services represents just two situations that require an exchange of information on technical data. Particularly with matters involving intricate details, and for reference purposes, the exchange should usually be in writing. When requesting information of this type, an important first step is to be clear in your own mind as to the exact information you wish to convey, or secure. With the facts clear in your own mind, plan the arrangement of your letter to make it easy for the reader to identify the specific information you are seeking. Tabulation is sometimes useful if a variety of points are being covered.

Inaccuracy or incompleteness can be costly to all concerned. This is especially true, for example, when seeking information from several vendors. In such cases your letter (if no formal request for proposal is prepared) should clearly stipulate the responsibilities of both your organization (the customer/user) and the vendor or contractor. In some situations your reason for writing may be to secure a variation or improvement in original specifications or some other special consideration brought about by an unforeseen development. Here, in addition to being specific about what you want done and why it is important, you will also want to be persuasive in your reasoning.

Other letters of this type might contain critical comments regarding technical data provided by a colleague or involve a line of reasoning that will help you secure greater corporate participation in a particular project or procedure. Inquiries regarding routine or relatively uncomplicated information should be as brief as possible, consistent with courtesy and clarity. Regardless of the circumstances, this chapter will help you cope with these letter- and memo-writing problems by providing tested samples that can quickly be adapted to your precise needs.

● *ACQUIRING DATA ON NEW SYSTEM*

General Guidelines

As illustrated by Letter 2-1, an inquiry regarding information on products or services should only be as long as it needs to be. The exact data required should be spelled out clearly and concisely, particularly in cases involving an initial inquiry. When the subject involves new products and/or services, don't suggest that a purchase is likely to be made in the near future if that is not a realistic possibility. Convey the impression you are searching for essential facts that could serve as the basis for such a decision. It is unlikely that any first inquiry will result in all the facts you need, and the door should be left open for follow-up questions based on the response to your initial inquiry. If it is imperative that you have the information by a specific date, indicate this. Otherwise, simply express your appreciation for an "early response."

Letter 2-1

TACOMA COMPANY INC.
1200 Forrest St.
Boulders, Illinois 00000

March 26, 19—

Richard's Electronic Laboratories, Inc.
One Industrial Avenue
Lowell, Massachusetts 00000

Gentlemen:

We are thinking of replacing our hardware with a more complete computer system in the near future and have a particular interest in getting detailed information about your "Star-Com" package.

Specifically, we'd like detailed information on its capabilities, price, maintenance requirements, costs, and its modularity as well as data on the documentation that accompanies the system, with an indication of how much training you would provide members of our staff. If there is a cost for training staff members, please indicate this.

I would appreciate an early response to this request.

Yours truly,

Alternate Approaches

(a)   We are designing several piping systems that will require extensive use of stainless steel pipes, fittings, and valves. Specifications call for Type 316 throughout, in a wide range of sizes (see enclosed listing), capable of withstanding pressures up to 3,000 psi. The data sheet with this letter indicates quantities we're considering, along with projected dates for delivery. As you will see, we expect to complete a test system by April 5 and would need the first delivery on April 12. After your review of enclosed data, please send detailed answers to the listed questions no later than March 20. Our engineer in charge of this project, Richard Benton, will get in touch with you shortly after we receive your response.

(b)   How soon would you be able to deliver 3,000 predrilled circuit boards of the same size, design, and quality as reflected by the enclosed sample? We anticipate a need to change suppliers, and your catalog does not carry price information on this particular type of board. If available, and in addition to your response regarding delivery date, please send design specs and price data on the full range of circuit boards you carry in stock. Provided the enclosed sample cannot be duplicated by any of the standard boards you carry, what would the necessary modifications cost in quantities of 3,000, 4,000, and 5,000? Our next production run on this product is scheduled for May 12, and we'd like to make a decision no later than May 1. Please let me have your response to these queries as soon as possible.

(c)   The speech synthesizer IC we're presently using is the SP0256-AL2. Circuitry changes being contemplated would require corresponding adjustments in other parts of this circuit (see enclosed spec sheet and a description of revisions planned). Following these changes we would be interested in making quantity purchases of the text-to-speech IC (CTS256-AL2) which I understand you have in stock. If you also carry 10-MHz crystals, we would need these too, in the same quantities. Please send detailed information on both items, along with price data on quantities of 1,000 to

10,000 in increments of 1,000. I would appreciate having this information within the next two weeks. Your response should be sent to my attention. If you have questions regarding the design changes, call the engineer in charge, Donald Clayton, at (201) 664-5795.

● *INFORMING STAFF MEMBERS OF CHANGES
    IN PROCEDURE*

General Guidelines

   Memo 2-2 deals with the type of situation that is frequently covered during discussions with staff members. Yet, as pointed out earlier, important procedural changes should usually be made part of the written record. This serves two purposes: greater thought can be given to spelling out the essential details, thus minimizing the risk of leaving out an important factor, and those involved can refer to it when implementing or communicating the change. The primary rules here can be summarized briefly.
   Give the pertinent information, making certain the primary details are spelled out clearly. If the objective in bringing about the change is not implicit in the memo or letter, make sure you put this in sharp focus. When the action will affect other departments, see that all concerned receive appropriate notification before the change is put into effect.

Memo 2-2

February 9, 19—

TO: Donald Axwell

FR: Bill Murray

Don:

As of March 15, 19—, the metal planer, operation no. 0330, will no longer be available in the Alderwild Machine Shop. We expect to install and begin operation of a new planer either on or before the March 15 date. This may lead to adjustments in the

```
fabrication project you requested on March 8, but if this re-
sults in any dimensional changes we will, of course, contact you
before going ahead with the job.
```

```
In any event, please note this when scheduling future work for
this shop.
```

```
Bill
```

Alternate Approaches

(a) I'm glad to tell you that your concern regarding difficulties with the metal planer has resulted in action that should please you and your staff. Your recommendation that we replace the planer in the Alderwild Machine Shop has been approved. As indicated in the details that follow, we expect you will be able to apply this new equipment to operation no. 0330 in the near future. Please make certain the procedural changes involved and the transferral dates indicated are made known to appropriate members of your staff by March 10.

(b) This memo deals with adjustments in the fabrication project (#0330) you requested on March 8. It has been decided to purchase three new lathes within the next month or so. The exact date of installation and other related details will follow shortly, but you should immediately prepare your staff for this changeover and send me a copy of your written communication to them. There is no way of knowing at this point whether we will need to make any changes in peripheral equipment, but if this becomes necessary, we will of course contact you before any final decision is made.

(c) Don, we will soon need your assistance regarding a new hydraulic press that has been purchased for use on, among other things, the fabrication project you requested on March 8. The bulletin attached to this memo will acquaint you with details regarding the actual installation and its location in the Alderwild Machine Shop. What I'd like you to do is meet with our staff prior to the installation so we can determine what effect this will have on the #0330 fabrication job. Some adjustments may be necessary in the production schedule because of dimensional changes that may become necessary. We're not certain at this stage whether this will be the case, but even if it isn't, I'd still like you to join us for the benefit of your suggestions on the transition phase. You will also want to notify your people of this, and if you like, feel free to send them a copy of the enclosed bulletin.

● *INVITING CONTRACTORS TO SUBMIT BIDS*

General Guidelines

Many corporations send out a detailed request for proposal (RFP) or a request for quotation (RFQ) with a cover letter when planning to purchase or rent equipment and/or contract for services. Memo 2-3 is an example of a cover letter that accompanies the specifications data provided to the contractor. If the product or service does not warrant the issuance of a formal RFP, the requirements can simply be stated briefly in a letter. In the latter case, and depending on the degree of urgency, telephone inquiries can be made, although written requests reduce the possibility of misunderstandings and needless discussions at a later date. In those instances when bids are requested and received by telephone, written confirmation should be secured as soon as possible. In all cases, it is important to state clearly and completely the information you want the vendor to provide and the date by which you must have the information requested.

Memo 2-3

WESTWOOD STEEL COMPANY
412 Ames Street
Allentown, Pennsylvania 00000

March 26, 19—

TO: (Contracting Firm)

FR: Department of Engineering, Purchases and
      Contract Services Section

RE: Request for Proposals Dated April 8, 19—,
      For a Computer Network Installation

PURPOSE: Westwood Steel Company desires to contract with an independent firm for installation of a network system to process, validate, verify, and interpret research data compiled by the

Product Manufacturing Division. Complete specifications are attached.

TIME: The sealed proposals should be sent under separate cover and subject to conditions stated in the enclosed RFP. The deadline for submission of all proposals will be 2:00 P.M., May 28, 19—, for furnishing the services described here. Procedures and requested data contained in the separate guidelines attached must be followed without exception.

NOTE: Indicate your firm's name and address and the words "Cost Proposal" or "Technical Proposal" on the front of each sealed envelope, along with the date of submission of your proposal.

Alternate Approaches

(a)   In addition to the detailed specifications sheet attached, we have also enclosed a rendering of design #40322. This illustrates the overhead walkway from our laboratory to the adjacent office building. We would like your quote in triplicate showing the cost of furnishing this walkway, ready for installation but not installed, including one coat of primer paint. While the dimensions indicated in the specifications are reasonably firm, we reserve the right to make minor modifications prior to the actual installation, which will be handled by our own staff. It is important that the deadline indicated be adhered to, and if you have questions, please call our engineer in charge of this project, Mr. Robert Farnsworth, at (201) 567-8900.

(b)   The attached RFP is self-explanatory, and we would like your bid on the items listed no later than March 15, 19—. Per our earlier discussion, if you wish to broaden your response to include additional services relevant to the project indicated, we have no objection to this, but it should be treated separately. Please bear in mind that the items listed would need to be shipped to the seven branch offices indicated, and shipping charges should be included. We would have preferred to give you more time in developing your bid, but delays encountered with suppliers of other equipment have caused us to accelerate our plans for completion of this project. After you have completed your computations, please call me with the estimates requested, and then send a written copy of your quotation to my assistant, Ms Ellen Fairchild.

(c) As indicated by the enclosed request and related specifications, American Steel Company is seeking bids for a network computer system to serve our six branch offices in the Northeast. We have some preferences as to the hardware to be used, but wish to base final decisions on your recommendations. The training of our employees is a particularly important aspect, and the data accompanying your bid should place special emphasis on this. Both the hardware and software presently in use by our firm are indicated on the separate list enclosed with this letter. Our timetable for completion of the installation is indicated, and we will need your detailed response no later than March 12. If you require additional information prior to sending your proposal, please submit your questions in writing to me with a copy to the head of our Data Processing Department, Mr. Robert Irving.

● *GAINING ASSISTANCE IN TRAINING PROGRAM*

General Guidelines

Letter 2-4 will be helpful when you invite colleagues to share their particular expertise with one or more members of your staff. This might relate to either formal or informal training programs that have been established. Your chances of gaining this type of cooperation will not be improved by taking an apologetic approach. Make no excuses. A letter that includes a statement such as "I know you're very busy and I hate to ask you this, but maybe you could find time to share some of your experiences with our staff" is likely to draw the response, "You're right. I don't have the time." There is an implicit endorsement of the reader's qualifications in a letter of this type. You are recognizing the person's special skill and knowledge in a particular subject area. You want to acquaint others with his or her skill, and believe they too will be impressed—and also aided— by the experience. This is no time for vague generalities concerning your reason for the request, and certainly no time for omitting important details regarding date, time, and place for the meeting. Convey too, provided it is realistic to do so, your willingness to reciprocate at some appropriate point in the future.

Letter 2-4

GENEVA MANUFACTURING CORPORATION
P.O. Box 629
North Cumberland, Maine 00000

January 2, 1987

Mr. William Blackwell
Engineering Consultant
402 Main Street
Bangor, Maine 00000

Dear Mr. Blackwell:

I am aware of your reputation as an outstanding engineer in the field of fiber optics. Would you be willing to spend a little time with three recent graduates of MIT in an effort to share some of your experiences in this field?

Our local chapter of the Northeast Engineering Association will have its monthly meeting on February 22 at 7:30 P.M. at the Hotel Jefferson on Bennington Avenue—which is just three blocks from our plant. The three graduates I mentioned, along with our entire engineering staff, will be in attendance. They would be delighted to join with me in meeting you prior to the program for a discussion of particular ways in which we would like to draw on your experience. I'm aware of several organizations that might be interested in your consulting services, and we can also cover this during our discussion.

Please let me know if you can join us on February 22.

Thanks very much, Mr. Blackwell.

Cordially,

Lester Dawson
Vice-President, Engineering

Alternate Approaches

(a)   An interesting question came up at one of our most recent staff meetings. Coincidentally, it led to the mention of your name. More than a "mention" really, because we spent a few moments talking about the special skills you've developed in the field of fiber optics. A few months prior to this meeting we employed two recent graduates from MIT, and it was one of the graduates who remarked about your outstanding experience in this field. The result of all this is twofold: we'd like to extend a warm invitation to you to attend our Northeast Engineering Association meeting on February 22, at 7:30 P.M., to be held at the Hotel Winthrop in North Cumberland. Second, our staff would like to share some data recently uncovered during their research in this area.

(b)   Your latest paper on fiber optics has just been reviewed by our staff and, to put it mildly, we're impressed. We find it extraordinary that so much of the research you're conducting represents areas we are now entering, principally because of two products that are now on our drawing board. Not content with just one coincidence, there is another one that compels me to send this invitation to you, that is, the fact that we have just established the basis for a training program dealing with current developments in the fiber optics field. I have enclosed a copy of the agenda we'll cover during the four sessions indicated. For these reasons, we sincerely hope you'll be able to join us in a discussion at one of the meetings, on whatever date that is most convenient for you. In addition to the significant contribution this would make to our training program, it seems likely that the resulting two-way communications would benefit all concerned.

(c)   Your considerable experience in fiber optics demonstrates an awareness that we closely identify with. That is, the need to capitalize more fully on known facts in developing practical applications in this field. We have made a strong commitment to this, as illustrated by the training program described in the enclosed bulletin. Products we now have underway further exemplify this commitment. Considering the things we have in common, it appears a mutual exchange of ideas and concepts could prove both interesting and beneficial to all concerned. With this in mind, we'd like to welcome you as a participant in one of our staff meetings. Some of the ideas and plans we now have under consideration (the bulletin describes only a few) may contribute to your research endeavors. And, conversely, your observations on many of these topics would certainly be of great interest to us.

● *INVITING COLLEAGUE TO TECHNICAL SEMINAR*

General Guidelines

It goes with the engineering territory to participate in seminars dealing with your area of specialty, either as a representative of your organization or as an individual. Memo 2-5 will be helpful on those occasions when you need to invite someone to attend as an active participant or speak at such a meeting. While these events can be an added drain on everyone's time, letters of this type can make a worthwhile contribution toward professional relationships with your colleagues and deserve closer attention than they often get. Whatever form you choose to follow, the important thing about an invitation of this type is to provide all the necessary facts. Check your letter to make certain you've covered such things as date, time, place; whether confirmation is necessary; your telephone number; the topic(s) to be covered; and whether you prefer the reader to indicate the primary subject of his or her remarks. In the last case, you should provide clues as to those areas you feel would be of particular interest to those attending the seminar.

Memo 2-5

March 2, 19—

TO: Helen Kiech

FR: George Anderson

Helen:

As you will note in the enclosed brochure, our seminar on "Women in Engineering" is scheduled for April 12, and I'm in the process of putting together the list of speakers. I believe that you, and perhaps a few of your colleagues, could make a solid contribution to this meeting.

In view of your contacts with other women in the field, I'd appreciate it if, in addition to your participation, you would

contact a few of these people about attending the seminar. It
seems likely that both the speakers and the audience would gain
from this experience. Provided you're interested, let's meet for
lunch next week, and I'll fill you in on all the details. Either
way, I'd appreciate hearing from you about this in the next few
days.

I do hope we're fortunate enough to have you with us at this
meeting, Helen.

George

Alternate Approaches

(a)   The Brookville Chapter of the Engineering Association is pleased to
extend this invitation to you to attend our forthcoming seminar on
"Women in Engineering." The enclosed brochure provides background
data. However, this is more than an invitation to attend. We sincerely hope
you will agree to be one of the panel members at this event and, if at all
possible, suggest two or three of your colleagues who might also like to
participate in the meeting. You may be certain our membership would be
delighted to have you with us, particularly because of the research you've
recently conducted in the field of robotics.

(b)   As indicated by the enclosed bulletin, the Northeast Engineering
Association has scheduled a one-day seminar on "Women in Engineering,"
and it is my pleasure to extend this invitation to you to attend as one of the
participants at this meeting. Because of your excellent article on telecommunications in last month's journal, we would be most interested in your
recent experiences in this field. It is a subject of considerable interest to our
membership, and you may be sure of a warm reception by others who
attend this meeting. Normally, we have an attendance of 100 to 125 engineers at these meetings, and we do hope you'll be able to join us.

(c)   The subject of "Women in Engineering" will be covered during the
next regional NEA meeting. Although several presentations will be made
by professionals in other engineering disciplines (see the agenda enclosed),
we are particularly anxious to have someone with your experience cover
recent developments in the superconductor field. Would you give us the
pleasure of having you as a participant at this meeting? We would be
pleased to have you, and if you have colleagues who would like to attend,
they would be welcome too. The enclosed bulletin provides the essential

details. Because of the need to include a list of attendees in our newsletter, would you let me know by the end of next week?

● *SECURING RESPONSE TO QUESTIONNAIRE*

General Guidelines

   Memo 2-6 provides an effective way to secure information on a broad range of subjects. The key to a successful letter of this type is clarity, and one factor that helps to ensure clarity is brevity. Too often, adding supposedly clarifying details becomes a distraction to the reader. If, for example, you wanted to preface your request with a statement such as "Starting in April, send your responses to the monthly status questionnaire to J. C. Henning instead of M. R. Dalton," an explanation of when Henning replaced Dalton, whether this is temporary or permanent, Henning's background and qualifications, whether Dalton quit or retired, and your regrets or congratulations are of no importance in getting the questionnaire rerouted. Repetition should also be eliminated. If your draft of a covering letter includes questions that are listed in questionnaire, omit them from the letter. You also need to explain the importance of answers and responses provided by the reader. Few people enjoy responding to questionnaires, although if properly designed, they can draw specific information more quickly than can many other forms of inquiry.

Memo 2-6

March 21, 19—

TO: Bill Carson

FR: Ellen Kern

Bill:

The state of California Department of Water Resources has asked for our help in an important survey of industrial water use. The objective is to obtain information on the specific purposes,

location, and amount of water used by the Waterford plant. This
will help develop realistic plans for more effective water management and resource development.

The state's last survey, in 19—, revealed a statewide industrial
water requirement of about 500,000 acre feet, or about 20 percent
of total urban use. Apparently, that percentage figure has changed
because of increased awareness of the need for water conservation
and because of the increased water cycling and reuse by industries
such as ours. The 19— drought emphasized the potential economic
threat of a water shortage. Today's demands on existing supplies
indicate that shortages can still occur unless we maintain and
strengthen our present procedures, while developing additional
safeguards for the future.

We can probably provide much of the information they need by
drawing from our 19— year-end reports. As you will note on the
enclosed form much of the data are already available to our various departments, and if there's additional information you need
from the headquarters office, please contact me directly. I
would like you to return the enclosed form to me by May 15, 19—.

Alternate Approaches

(a)   In developing plans for the design and production of the X-400
monitor, we have generated a number of key questions that need to be
answered by appropriate staff supervisors. Once your answers have been
compiled and coordinated, we'll be able to develop a sharper focus on how
best to proceed with this project. As you know, our target date for completion of the plans is April 21. Everyone receiving this brief questionnaire has
a vested interest in making certain we meet that deadline. While a concerted effort has been made to list only the most essential questions, don't
hesitate to add any additional points or suggestions you feel are relevant.
We've "tested" the response time required, and this indicates it should only
take you about 20 minutes to fill this out. It will be time well spent. The
eventual results will benefit your department as well as ours, and we're
convinced that the thought and effort you put into this now will pay off
later.

(b)   We have just received the enclosed questionnaire from the state
Department of Industrial Resources. As you will note, participation in this
survey is not voluntary, it's compulsory. The consolation is that we will be

able to benefit from the results just as much as the state will. Actually, the research itself is being conducted for reasons that are important to all of us, so I would like you to give this your best effort. You will notice a concluding section in the questionnaire labeled "Comments." This provides an opportunity to register our concern regarding recent state tax law changes discussed at our last meeting. Please fill out the questionnaire, but before you complete the "Comments" section, review the attached memo from our comptroller. Then, include whatever thoughts you have on this subject in that final section of the questionnaire.

(c) Our corporate membership in the Engineering Association represents a mutually beneficial relationship that has lasted for a good many years. Occasionally, however, many of us have remarked about the lack of two-way communication, and the feeling that the rank and file membership does not always have a sufficient voice in operational matters. Now, we have an opportunity to express our feelings on the subject. We have been invited to participate in a survey, as reflected by the attached questionnaire. There is only one copy, and our response should be coordinated so as to convey accurately our collective thinking on each point covered. Since you are one of our corporate representatives who attend the various meetings, seminars, and exhibits sponsored by the Association, we want to make certain you have a voice in the response to this questionnaire.

Please send me a memo that provides your overall view of the Association, with particular emphasis on constructive changes in procedures or policy you would recommend. Since we need to return the completed questionnaire by May 12, please let me have your memo no later than May 5.

- *CONVEYING CRITICAL COMMENTS ON RESEARCH DATA*

General Guidelines

While few engineers relish the assignment, it sometimes becomes necessary to deal with questionable data compiled by other colleagues. Letter 2-7 deals with one such situation. There will inevitably be other occasions when you disagree with an article, research paper, or other published material. Admittedly, writing effective letters that criticize, make an opposing point, or correct someone on a particular issue is not always an easy task. But if you feel the issue in question is an important one, it provides an opportunity to express your opinion in a constructive way. In a real sense, gaining cooperation in the correction of inaccurate technical data can be considered one of the engineer's important responsibilities.

Moreover, letters of this type involving the media, especially when your title and affiliation are indicated, can have a significant effect on those who read your comments. Clarifying or correcting misleading data places them in a better position to understand and/or evaluate the issue. It is especially important that letters of this type be well written, accurate, and crystal clear.

Letter 2-7

STARKTON MANUFACTURING COMPANY
Soya Avenue
Decatur, Illinois 00000

February 17, 19—

Boyston Technical Journal
1520 West Street
Chicago, Illinois 00000

Gentlemen:

The article on America's electronics industry in your January issue was a textbook example of the "creative" use of statistics. It reminded me of the claim made by an executive with the National Meat Institute that since Americans eat 19 billion hot dogs per year, that equates to 0.35 ounces of hot dog per person per day. The claim, of course, was based on the premise that every person in the United States eats some hot dog each day, an obvious fallacy. Using this type of reasoning, you could "prove" that potassium cyanide is not poisonous by asserting that a pound distributed evenly through the entire population of the entire United States would provide only about 2 micrograms per person. My doctor says that the lethal dose for humans is about 1,000 times that much. I wonder if that individual would willingly ingest the amount proportionate to his personal percentage of the annual U.S. hot dog consumption, just to support his position.

The article's assertion regarding TV sets, compact disks, and the general population was equally fallacious. Note the correct

(and documented) figures enclosed. I read and admire your journal regularly and would appreciate your cooperation in correcting this inaccuracy in a future issue.

Sincerely,

Matthew Gray

Alternate Approaches

(a)   It has been said that a statistician is a person who can draw a
mathematically precise line from an unwarranted assumption to a fore-
gone conclusion. I was reminded of that bit of whimsy when reading your
paper on OP-AMP technology. Actually, I was impressed with the comprehensive nature of your remarks, and perhaps this caused me to be more
sensitive to the basic omission noted. The fundamental information that
was lacking would have helped to make an informative presentation even
more helpful to the reader. Specifically, the basic definition of an operational amplifier is something every electronics engineer is aware of, but
that is not true of the audience you were writing to. Had you indicated,
in the beginning of your paper, that the Op-Amp is a high-gain, direct-
coupled amplifier that uses feedback for control of its response characteristics, this would have put your subsequent remarks in sharper focus for
the reader.

(b)   Your April 28 editorial on "Pensions and the Professional" contained several misleading elements which can only serve to prejudice and
mislead the public. It is indeed possible that in a period of sustained high
inflation, the retiree you described could earn more in retirement than he
once received in salary. The annuity, however, would have to be pegged to
inflation. This is more a commentary on inflation than a valid criticism of
indexing, because the primary purpose of indexing is to protect the buying
power of the original annuity—no more and no less. You also quote the
governor as saying that twice-a-year adjustments are an "easy target." We
would suggest that they have been made a very *visible* and unfair target by
editorials such as yours. Many of your other points were well taken, and
we appreciate the attention you are giving this general topic.

(c)   Your research paper on linear motors offered many interesting
insights that attracted the attention of our engineering staff. There was,
however, one element lacking that would have contributed a great deal of

practical value had it been included. That is, more detailed documentation of the reference works it was based on. The few footnotes were really
no substitute for this expanded documentation. Fortunately, the paper
reflected your continuing research in this area, and we hope you will
update this presentation at the earliest possible date. Provided you do, it
would enhance its value if you added a reference section that clearly
identified the sources. In turn, and provided you would like us to do so, we
would be glad to share our experiences in this area with you, particularly
those that have a direct relationship to the applications suggested in
your paper.

● *SECURING ASSISTANCE FROM ENGINEERING CONSULTANT*

General Guidelines

While Letter 2-8 will help secure the specific information requested,
there are other situations that might also require a letter of this type. The
purpose may relate to only one particular aspect of a project you're involved with, or the development of an overall program dealing with design,
production, and/or research problems. Regardless of the circumstances,
you need to convey the exact parameters of your request to secure a considered and appropriate response from the recipient. The situation is similar to
drawing on computer-based data; for example, the utility value of the response you receive depends on the information you convey. Put another
way, quality of output depends on quality of input. Clarity and completeness will help ensure that your letter results in the action and assistance you
are seeking. When writing to a specialist of this type, it is also important to
pinpoint the specific objectives of a particular assignment. Otherwise, the
consultant may feel inclined to broaden the scope of the activity you want
him to perform. Spelling out the limitations at the outset will help avoid
this problem.

Letter 2-8

MICRO ELECTRONICS CORPORATION
616 Ames Drive
Waycross, Georgia 00000

March 27, 19—

Mr. Peter Wainwright
Acme Engineering Services, Inc.
412 Benton Blvd.
Savannah, Georgia 00000

Dear Mr. Wainwright:

Because we have used your counseling services before and were
satisfied, we are again seeking your expert advice.

As you know, we have been using the XY4112 chip for our R12 amplifiers ever since you completed the circuit design for us in
January. It has been satisfactory, but we want to explore various alternatives because of plans to diversify two of our
product areas.

What we want from you is an evaluation and guidance on how other
chips might affect the performance of this solid-state circuitry, for example, BRA-400, BRB-464, and BRC-500. I assume
your research will also determine the effect this would have on
amplifier output (see enclosed schematic).

Before your specialists start work on this assignment, please
give me a written estimate on potential costs and the time frame
involved in this project.

Sincerely,

Alternate Approaches

(a) For a variety of reasons our chief radiation physicist at the clinic
has requested that we compile research data on developments in linear

accelerators in the past two years. I recall from past conversations with you that two leaders in this field are Varian Associates, Inc., and the ATC Medical Group in Ohio. There are undoubtedly others who possess up-to-date information that could be of practical value to us. We would like you to do this study, following the format illustrated by the attached chart. As you will note, the data requested include such topics as RF source, energy levels, dose rate, field size, and symmetry. Provided you are able to take on the assignment you should first review the chart carefully. It pinpoints the precise information we're seeking. Do you feel it's complete? Are there areas you feel we should add? Anything you feel should be omitted? Would this format simplify the collection and coordination of information? We'll also need firm cost and time estimates.

(b) Our next assignment for you involves the circled portion of the schematic diagram I've enclosed with this letter. This FM demodulator circuit has been giving us a problem with the prototype being developed for the X-400. A working model of the chassis is being sent separately. You've had recent experience with gated-beam detectors (remember the X-30B?), and I think you could probably determine the necessary modification without too much trouble. If we had enough time and fewer other projects, we could probably figure this out, but it's a rush assignment. We want to get the X-400 into production as soon as possible. Can you give this a top priority? It would be appreciated. Give me a call as soon as you've gone over this and let me know your fee—and how long you think it will take you to complete the assignment.

(c) We're not pleased with the quality control program currently in use at our plant. An analysis of client complaints over the past year (see enclosed listing) convinces us we need to do a lot better in establishing, and maintaining, a program that ensures fewer product defects. Since your brochure reflects broad experience in this area, including your position as director of quality control for Apex Corporation, we may request you to conduct a survey for us. The purpose would be to compile research on programs now in effect at comparable manufacturing firms across the country. I have also enclosed a list of the specific types of data this survey would be designed to uncover. Any comments or suggestions you have on this listing would be welcome. Following your consideration of this inquiry, please let me have a letter detailing your thoughts on the assignment. Since this research could have an effect on budgetary matters presently being considered, we'd like the survey completed as soon as possible. Please include in your letter an estimate of the time and cost involved in conducting this survey.

● *ACQUIRING TECHNICAL INFORMATION FOR USE IN SAFETY PROCEDURES*

General Guidelines

The writer of Memo 2-9 realized that providing extensive technical data in response to a request may take a considerable amount of time. It may also reach the recipient just when he or she can least afford the expenditure of that time. Thus, it becomes important to offer a persuasive and convincing explanation of the reason for the request. There are a few questions you might ask yourself before writing such a letter, for example, what reasoning will appeal to the reader? and how are your needs tied to the reader's own interests? A request of this nature is more willingly granted if the practical psychology of give-and-take is recognized. The reader may ask, "What's in it for me?" Whenever possible, offer something in return. In the case of the sample letter, the answer becomes obvious. The identification of all potential hazards associated with the chemicals referred to is a factor that will determine the continued use of these chemicals. It is clearly in the best interests of the supplier to provide comprehensive and reliable answers to the questions being asked. Never end such a letter with a statement like, "Thanking you in advance" or the shorter "Thanks for your help." These phrases leave the reader with the impression that the writer is terminating his or her interest in the request, and the reader is left to struggle on alone. Rather than closing with a curt "Thank you," end on a courteous note that more fully conveys appreciation for the reader's willingness to cooperate.

Memo 2-9

March 21, 19—

TO: John Chapple, Chemical Supply Division

FR: Don Brinkerhof, Production Department

In order to ensure the safety and health of our staff employees and comply with current government regulations, we must update

our records on hazards associated with the chemicals currently
used in the Chemron product line.

All of the questions on the enclosed SAFETY ADVICE are important.
We would like a response that leaves no spaces blank. Answer each
question as completely as possible. Your comments will be reviewed carefully by the Management and Safety committees. If
further information is required, we will contact you for the
specific data needed. Any future changes in composition of these
chemicals should be reported promptly with a revised SAFETY
ADVICE data sheet.

Your assistance is appreciated. We look forward to receiving
your response no later than April 15.

Alternate Approaches

(a)   I have attached Alex Smith's proposals regarding operations in
Plant 2. You will note he has covered two major areas: simplification of
our die usage calculations and safety procedures that should be installed
immediately. His recommendations regarding die usage are sound, and I
have, without exception, approved each one. The material regarding
safety procedures, however, is something else. They appear reasonable,
but you're the expert in this area, and I'll rely on your evaluation of the list
provided by Alex. Please go over each recommendation carefully, discuss
this with your appropriate people, and then let me have your detailed
response. Since this relates so closely to the welfare of our employees *and*
our compliance with state and federal safety regulations, I'll need your
response no later than April 20. If you have questions that need to be
answered before beginning your evaluation, please call me on Ext. 2112.
I'm glad you're the one who will be deciding on this aspect, Tom. Alex has
been told not to implement any safety procedural changes until we have
your considered response.

(b)   Because of the recent increase in our accident frequency rate, the
district manager has asked for an intensive effort to revamp and improve
our safety program. As one of the key supervisors directly responsible for
employee safety, I need your best thinking and full support applied to the
attached list of recommendations. Do not consider this list all inclusive.
There may be additional areas you feel should be covered. If so, please add
them. While every item on the list is important, I want you to pay particular
attention to the section in which you will provide an analysis of on-the-job

accidents in your department over the past year. You will need to supply a brief description of the cause, the resulting injury, and action that has been taken to avoid a repetition. Our obligation here is not only to the employees and the corporation, but we must also be aware of government regulations that need to be observed. To make certain that everyone is aware of our concern, and the resulting investigation, this memo is being posted on all departmental bulletin boards. As supervisor, you should discuss the matter with your staff. We want to make certain their comments and suggestions are carefully considered. The final result of this survey should be a substantially improved safety program, to become effective no later than May 15. Thus, I would appreciate your response no later than May 1.

(c) I know this is a busy season for you, but "safety first" is more than a common expression. For us, it has recently become an imperative. As indicated during our telephone conversation, the 12-volt power supplies shipped to us in March have caused minor electrical shocks to three of our employees. They were "minor" only in the sense they did not cause serious injury, but the obvious question is indeed a serious one. To supply you with all the relevant facts, I have asked the supervisors to provide a written description of each incident. Their responses are enclosed with this letter, along with excerpts from our shop manual showing how, when and for what purpose we use the power supplies. I am providing these details for three reasons. First, we've done business together a long time, and I'd like to see this continue. Second, it is possible that some of our procedures may need to be altered when using this equipment. Finally, it may become necessary to purchase additional power supplies with a variable output up to 30 volts. At the earliest possible date, please let me have your recommendations regarding the 12-volt units and detailed specifications on dual-tracking DC power supplies with output ranges up to 30 V DC.

● INVITING PARTICIPATION IN RESEARCH PROJECT

General Guidelines

Letter 2-10 will help simplify an assignment that is not uncommon in the engineering community, specifically, the need to draw on the experience of appropriate professionals in compiling useful research—research that will be of benefit to others both in and outside your organization. It takes a carefully drawn and convincing letter to persuade a busy individual to devote some of that time to you. The sum of such a letter is comprised of a few essential parts. One key is the need to be specific about the research project itself, with emphasis on *why* the project is an important one. Underscore the knowledge the reader has that you need. Don't be vague or hesitant when making the request. Beating around the bush is, in itself, a waste of time and may imply the same would be true of the research you are writing about. Emphasize your recognition of the individual's particular skills and the importance of their participation in the project.

Letter 2-10

PREMIER ELECTRICAL CO.
455 Gainsboro Avenue
Raleigh, North Carolina 00000

March 3, 19—

Ms Eleanor Stevens
Design Engineer
Albin Products, Inc.
Chicago, Illinois 00000

Dear Ms Stevens:

As a professional engineer in our community, you may be aware of the important research work being done by the League for Engineering Education.

This year we have expanded our operations and have undertaken additional responsibilities never before attempted by a similar group. A description of the current research project is enclosed.

Because of this ambitious effort for the benefit of all engineers in the area, there is an urgent need to draw on the talents and abilities of people like yourself. Could you contribute a little of your time to this worthwhile endeavor? Not only the League but the entire engineering community would be most grateful. The more capable people we can bring in on the project, the less time it will take individual members to complete their assignments.

In any event, so that we can make appropriate plans, I'd appreciate discussing this with you in more detail. At your convenience,
would you give me a call in the next week or so?

Thanks very much, Ms Stevens.

Cordially,

Alternate Approaches

(a) We all know how tragic the plight is of those stricken with
Alzheimer's disease, and we sympathize with the victims. Many of us try to
provide assistance in one form or another, yet how many individuals can
say they are doing as much as they can do? Professionals in the medical
community have much more than a vested interest in the problem and are
applying their various skills toward its solution. There is one area, however,
that all of us draw from, but sometimes do not contribute enough to. That
area involves research, and our local branch of the _____ Society, with
your invaluable help, will try to do something constructive about this. To
assist in the effort, we're contacting five leading professionals in the field.
You are one of them. Our objective and a general description of the project
is described in the enclosed literature. We would be most pleased if you
would join us in this collective effort.

(b) In developing the analytical study of ICs described in the attached
newsletter, my staff tells me there is someone important missing from our
plans. That someone is you. If we're going to do the type of job on this that
will benefit a broad cross section of electronics specialists, we need an
individual with the extraordinary range of experience you've had in this
area. I hope you will agree that this particular study would represent a
unique compilation of data that could prove useful to all of us. Because
this is a collective effort, it is unlikely that any one member of the team
would be overburdened by his or her particular assignment. As you will
note, a relatively thin (but crucial) slice of the research would be assigned
to you. As director of the League of Electronics Engineers, I extend this
invitation for you to join us. We would welcome your participation and
would be glad to set a date for our first meeting that would be convenient
for you.

(c) There are many issues that must be included in our agenda for the
coming months. The Association has grown rapidly over the past several
years, and we need to take stock of the manpower resources available to

us because of this growth. One practical way to do this is to develop a statistical profile of the membership. This could generate useful research data that would be invaluable when drawing up plans not only for next year's activities, but for future years as well. I cannot think of anyone in the organization who would be better qualified to quarterback this project than you. Your excellent professional background and gracious manner would help draw the necessary cooperation from others. It need not be a "rush" assignment, and our schedule could be built around your availability for the few meetings that would be necessary. I would be glad to work with you in establishing a support group to help you with the project.

● *INFORMING STAFF REGARDING INSTALLATION OF NEW EQUIPMENT*

General Guidelines

Memo 2-11 illustrates an effective approach to memos that convey essential information to staff members. When issuing communications of this type, there are two things in particular to keep in mind. The first, making the announcement clear, is obvious. The second element is embodied in the style and approach of your memo or letter. You will want to avoid a brusque, dictatorial tone. Presumably, you are making the announcement to inform and to elicit appropriate responses from those who will be affected by the information you're conveying. This does not suggest that you need to take a condescending or plaintive approach. It is simply a question of presenting the communication in direct, straightforward terms. If the reason is not clear as to why staff members should respond in a certain way, make it clear. In the case of Memo 2-11, the fact no electrical power would be available during the period indicated clearly conveys the basic reason for issuing the memo.

Memo 2-11

March 9, 19—

TO: William Atwood

FR: John Barnes

Bill:

A new 1,000-KVA electric transformer will be installed in the
Sitcom Plant during May 19—.

As part of the installation procedure, Worthington Electric
requires eight hours of downtime on all existing electrical
systems.

The downtime has been scheduled for Saturday, March 13, 19—.
There will be no electrical power in the Sitcom Plant from 8 A.M.
on March 13 to 4 P.M. on that day. Please notify all appropriate
personnel of this action. If you have any questions, contact me
immediately on Ext. 2112.

John

Alternate Approaches

(a)   We have come up with a way to avoid repetition of the type of theft
and resulting damage that occurred in the lab on March 10. It has been
decided to add an ultrasonic motion-sensing alarm to our security system.
An automatic timing device will turn on the system during off-hours, and
your supervisor will acquaint you with the operating procedure for this new
device. This addition to our system will be important to all of us, and I'd
like you to become thoroughly familiar with its operation. To avoid any
disruption in your individual assignments, the system will be installed on
Saturday, March 24. Please avoid any overtime work on that day. If you
must be in the laboratory on March 24, please let me know immediately.
We'll have a staff meeting on March 23 at 10 A.M. to discuss this, but if you
have questions in the meantime, call me on Ext. 2444.

(b)   Eight new portable computers will be delivered to my office next week for distribution to staff supervisors. They offer 16-bit technology, and each one contains a 3½″ disk drive, an internal direct-connect modem, and an 80×16 high-contrast LCD display. We're having an outside firm convert our software for use on these units, and I expect the conversion to be completed within the next two weeks. Operating manuals and appropriate documentation will, of course, be distributed with each computer. Please make certain you go over this material carefully before using the unit. On the day following delivery, we'll have a meeting in my office. Distribution of additional software (see attached list) will be made at that time.

(c)   Because of the relatively exotic circuitry involved in some of our newer products, we've decided to replace a number of the measurement and testing devices now being used in the shop. Specifically, all VOMs and VTVMs will be replaced by new benchtop digital multimeters. Fortunately, each new unit will have a separate diode-check mode for quick open/short/leakage tests of semiconductor junctions. In addition, they have a built-in memory function that stores minimum and maximum values of a changing input and displays them at the touch of a button. I have arranged to get a supply of the operating manuals and expect delivery in the next two or three days. The units themselves will not be available until the end of this month. When I know the exact delivery date, I'll pass this along to you. I hope some of you will have ideas on how best to dispose of the older units and would welcome your suggestions on this.

● *NOMINATING COLLEAGUE FOR TECHNICAL PROJECT ASSIGNMENT*

General Guidelines

Letter 2-12 illustrates one approach to take when voluntarily recommending someone for a particular assignment. This action places a clearcut obligation on the part of the writer. There is little point in nominating or suggesting someone unless you are convinced of the person's qualifications, ability, and integrity. Naturally, the recommendation itself should include a realistic endorsement of the person you're suggesting. Ideally, the praise should be based on your own personal experience with the individual, not on secondhand information. Inject a positive tone in your letter, providing only the necessary key facts. On the assumption that one usually makes recommendations of this type to established business contacts, as opposed to strangers, the tone should also be friendly and somewhat informal. If you are not certain the recipient needs the particular expertise possessed by the person you're recommending, state this. It then becomes a matter of sending the recommendation along "in the event you need a person skilled in this subject area at some point in the future."

Letter 2-12

TRANSPACIFIC AIR FREIGHT
435 Magnolia Avenue
South San Francisco, California 00000

March 3, 19—

Mr. Harold Wright
Acme Chemical Products
Denver, Colorado 00000

Dear Harry:

When we had lunch together a few weeks ago, I mentioned Jack Hunt to you, the project management consultant who did such an excellent job for us earlier this year. Since that time he has decided to relocate to your part of the country, Colorado Springs. His business card is enclosed.

I don't know whether you're interested in project management consulting at this point, but I do feel Jack could offer some novel and productive ways to make project management at your facility more cost efficient and profitable. Our past experience with him convinces me that he really knows his field.

Harry, why not get together for a detailed discussion with Jack? It won't cost anything, and my guess is that you and your firm may benefit from it. If you decide to do this, I'd be interested in the outcome.

Sincerely,

Alternate Approaches

(a)   Jim Wainwright told me recently that you were looking for a programmer, and I know of someone you might be interested in. Marsha Kennedy was employed at our DP center as an applications programmer for the past three years. Her husband has been transferred to your city, and this led to Marsha's resignation. We were pleased with her productivity and efficiency during the entire time she spent with our organization, and I would not hesitate to recommend her to you. She approached each assignment as a professional and got along well with both supervisors and her colleagues. I've enclosed a copy of the resume that became part of her file when she first applied for a position here. You will note that her previous experience included four years as a programmer with the Walton Corporation. If you have specific questions regarding her assignments with us, don't hesitate to call Tom Kern, her former supervisor in our DP Department. His number is 234-5678.

(b)   Paul Bell has been a key member of our consulting group for the past five years, and I can unequivocally attest to his product design ability. Paul has also been active in the development of technical manuals for the product lines we're associated with and has done a superb job in this area too. He told me of his recent conversations with you, and quite frankly, I have mixed feelings in writing this letter. We definitely would not want to lose Paul, but neither do I want to stand in the way of his advancement. From what Paul has told me, I can only gather that the opportunity you have in mind is one that offers quite a challenge for him. He is most interested, and regardless of the eventual decision, I simply want to let both you and Paul know that we value him as one of our most capable and dedicated employees. If I can be of additional help to Paul and yourself in bringing about a decision in the matter, you have only to ask.

(c)   This letter will introduce you to Alan Barry, a fine mechanical engineer I've had the pleasure of working with on various assignments over the past four years. Because of his wife's health, Alan must relocate to Arizona, and I told him I would try to pave the way for an interview with you. I don't believe you would regret spending some time with him. Alan has, on several occasions, been asked to solve the type of hydraulic press problems your plant is faced with from time to time and has come up with ingenious solutions that not only corrected the problems but saved money as well. He is a self-starter and, as you will see from his bio data, has excellent credentials. I've had a close working relationship with Alan since his senior year at Purdue and would not hesitate to recommend him to you.

If you have an appropriate position available now, fine. But if you don't, perhaps such a position will develop. With that thought in mind, would it be all right if I asked him to call you for an appointment?

- ## SECURING CORPORATE PARTICIPATION IN PRODUCT DISPLAY

General Guidelines

Letter 2-13 takes an informal approach to the matter of securing corporate involvement in technical product exhibits or displays. Such a letter can be an effective instrument for gaining goodwill as well as spreading useful information regarding product development. Again, it becomes essential to highlight the key facts concerning sponsorship, date, time, who will attend, and why it would be in the best interests of the recipient to participate. The last point may be obvious by the very nature of the exhibit, but should be reinforced. Wherever necessary, the letter should be accompanied by a bulletin or listing that provides additional information on others who are attending and/or exhibiting products. This addendum to the letter should also contain detailed information on fee arrangements if there is a charge for setting up product displays.

Letter 2-13

ABERNATHY ELECTRONIC DEVICES
Century Park
Brooks, Massachusetts 00000

October 7, 19—

Mr. Gerald E. Rogers
Richardson Software Corporation
820 Mark Street
Hartford, Connecticut 00000

Dear Gerry:

On Thursday, December 4, the local chapter of the Data Processing Management Group, in joint endeavor with the Association of

Electronic Technicians, is putting on a computer fair in the
Green Room of the Plaza Hotel. I would like to invite you to display your products in a special exhibit area being established
for this meeting.

The affair will start at 5:30 P.M. and will include several hardware and software vendors, demonstrating a variety of products
and software packages. Additional details are provided in the
enclosed brochure. We will also have Ross Glazer from the National Association for Computing Hardware speaking after dinner.

I do hope you can join us on December 4, Gerry. I believe you
would enjoy the presentations, the dinner, and the remarks of
Ross Glazer who is a very entertaining speaker. Would you please
let me know in the next few days whether you and your firm will
participate?

Sincerely,

Ross

Alternate Approaches

   (a)   The local chapter of the Design Engineer's Association will have
several product display booths at our annual meeting on March 10 beginning at 7:30 P.M. in the Madison Room at the Hilton Hotel in Tappan, New
York. We cordially invite you to select and exhibit several of your products
on this occasion. We expect over 100 professionals to attend, representing a
cross section of design engineers from various industries in this area. The
exhibits have always been one of the most popular features of our annual
meeting. Feedback from corporate representatives who have participated in
the past reflects the mutual benefit gained by both exhibitors and attendees.
Many of our members have a keen interest in your products, and I do hope
you will join us on March 10. The enclosed bulletin provides additional
details, and we would appreciate having your response no later than April 4.
   (b)   Your technical manuals have long been admired for their clarity and
emphasis on practical aspects. As a member of the National Organization
of Technical Writers, it is my pleasure to invite you to our next meeting on
June 11. The enclosed brochure will acquaint you with the details regarding this meeting, and we expect to have over 75 members in attendance.
Actually, we're not only inviting you to attend, but we'd be delighted if you

could bring along several "representatives" of your fine work in the form of manuals you have produced. A special area will be set aside in the meeting room so they can be displayed properly and in an attractive setting. When this matter was discussed at our last meeting, the membership was enthusiastic about the possibility of your participation. May we look forward to having you with us on June 11?

(c)   The subject of telecommunications will be featured at our next meeting. Fortunately, we've been able to secure four leading experts in this field who will make presentations to the membership on topics that are part of the cutting edge in this field. The enclosed data go into more detail about the subjects they will cover and will also give you an overview of our organization. Several members have expressed the feeling that a display of telecommunications products would add an excellent and very practical feature to this meeting. That leads me to this double-barreled invitation. First, we'd be pleased to have you as our guest at this meeting. Second, a display of selected telecommunications products from your organization would be of great interest to the membership. Would you like to explore this? It appears likely that such a display would offer mutual benefit, and I do hope you decide to investigate the possibilities. I'll wait a week or so to give you time to consider this, and then give you a call so we can discuss it in more detail.

● *ACQUIRING PRACTICAL INFORMATION ON USER EXPERIENCE*

General Guidelines

Letter 2-14 is a good example of a letter that blends both a personal and professional approach to securing useful data. One of the objectives here is to establish the type of cordial relationship that is conducive to a mutual sharing of information. Whether you wish to secure firsthand data from a personal acquaintance or total stranger, the same techniques of organizing and presenting facts will be used in a more formal business letter. As is true with practically all inquiries, you would do well to keep the reader uppermost in your mind. Write in a way that will quickly establish a common bond that reflects your mutual interest in the subject at hand. A letter, even a friendly note, is written to accomplish a purpose, and the key to this accomplishment is to think the way the reader does. Put yourself in his or her place and ask, "How would I react to this letter if I were the reader?"

Letter 2-14

WHITNEY CHEMICAL CORPORATION
854 Gary Boulevard
Philadelphia, Pennsylvania 00000

October 2, 19—

Mr. George Patterson
Royal Court Avenue
Chicago, Illinois 00000

Dear Mr. Patterson:

We have had the Adenhaur starch-making system in operation at our
plant for two months, and I understand you have been using this
system for the past three years.

I will be in Chicago during the week of November 19 and would
like very much to visit your firm to see your system in action.
We have found several interesting solutions to problems posed
during the conversion process, and perhaps an exchange of information would benefit us both.

I can arrange to visit your plant any day during the week of
November 19, but if this would be an inconvenient time period for
you, other arrangements can be made. Please let me hear from you
as to your preference.

Thanks very much, Mr. Patterson.

Cordially,

Arnold Bridges
Manager, Chemical Products Division

Alternate Approaches

   (a)   Our chief engineer tells me your firm has one of the best engineering staff manuals he has ever seen. I was especially pleased to hear this
because we are in the midst of changing many of our technical procedures

and policies. I don't need to tell you how complex and time consuming this can be, and perhaps your understanding of this will help determine your response to this request. It was my intention (and still is) to call you about this but wanted to send this note along before doing so. In essence, I'd be appreciative if we could get together for an hour or so next week to discuss the way you went about revising your manual of operations.

Our products are not competitive with yours, and we have no desire to get into anything that might be considered proprietary information. It is mainly your procedural approach in which I'm interested. In any event, I appreciate your consideration of this query. I'd be glad to have you as my guest at lunch, and perhaps when I call next week, we can set a date that's convenient for you.

(b) For some time now we've been considering the purchase of a T600 computer networking system, which I understand you've been using for the past two years. I suspect you went through much the same research study we're involved with, so you can probably understand my reason in writing to you for an appointment. Coincidentally, it will be necessary for me to travel to your city in the next few weeks. It would be great if we could get together while I'm there to share our mutual interest in this system. I would be glad to tell you of some of the things we've learned from our research, and I'm sure you could be helpful to us regarding the configuration on which we should focus. My schedule is flexible. I'll be glad to meet with you at your convenience. Could you let me have your response in the next week or so?

(c) Because of circuitry design changes, coupled with cost considerations, the CRTs we've been using will be phased out during the next few months. Our supplier tells us you recently made a switch to the 19ARL and have been very pleased with the results. That makes you an ideal person for us to talk to. Since we do not manufacture sets that compete with yours, I'm hoping you'll be willing to share your thoughts with me on the 19ARL. The schematic I've enclosed represents the new circuitry we'll be using. As an engineering colleague, what I'd like to do is spend a little time discussing your reasons for switching to this CRT, your experience with it, and whether you plan to continue using it in the future. As you know, we specialize in smaller sets, but I do have some interesting data on 19-inch and larger sets that I'd be glad to share with you. Please let me know when it would be convenient for you to join me at lunch so we can discuss this in more detail.

A WORD ON STYLE

Make Every Word Count

Writers on prose style always stress conciseness. Instead of General Bedford Forrest's "Get there fastest with the mostest men," the rhetoricians seem to say, "Get there fastest with the leastest words."

On the face of it, there is nothing wrong with that advice. One of the most prevalent sins in modern letter writing is wordiness, the use of two or more words when one would do. But the obvious solution—cut, cut, cut—may create more problems than it solves.

In writing anything, there is an optimum number of words, not a minimum number. An editorial genius will strike a point somewhere near the ideal figure. An average writer will either overshoot it or undershoot it.

There is no magic formula for concise writing. Some books on writing tell you to write the way you speak. It sounds reasonable, but just try doing it.

The best advice is this: Read your own writing. Think about it. Does it make sense to you? Would it make sense to you if you didn't already know the message? If you have two yeses to these questions, you have probably produced a letter that is as concise as it needs to be.

Chapter 3

LETTERS AND MEMOS RELATED TO PRODUCTS OR TECHNICAL SERVICES

The nature of the engineering profession requires many letters and memos to focus on a particular need or objective. Few, however, are more important than those that identify, and enhance, the benefits offered by a useful product or service. Some engineers do not relish being in the position of "selling" someone on the relative merits of their product, but in reality, they are often in the best position to understand and pinpoint the features that can help ensure the success of that product or service.

Many letters dealing with broad engineering concepts are not designed to produce immediate and/or discernible results over the short term. This is generally not the case with letters that pinpoint the benefits offered by a specific product. Written communications of this type should coincide with (and reinforce) the overall marketing effort. Contrary to many other types of letters or memos, the results can be more quickly determined. Consequently, letters that dovetail with the marketing aspects offer special challenges to the engineer. They need to stand out above the more routine forms of communication. When handled successfully, the rewards are tangible and gratifying. The outcome can be translated not only in terms of profits, but also in terms of personal satisfaction.

This chapter will help you write letters and memos that extend the engineering contribution to a level that not only supports, but advances the marketing effort.

● *CONVEYING DATA ON PRODUCT IMPROVEMENTS*

General Guidelines

Letter 3–1 reflects one of the more common situations that provide marketing-related opportunities. For example, a thoughtful letter in response to a person's interest in a seminar (whether or not they actually attend) opens the door to a more productive dialogue. The situation has several variations. They may have visited your booth at a product display, possibly contacted you or someone on your staff for product information, or asked a technical question about one aspect of your product. Your response should clearly and quickly explain the reason for your letter. Summarize the major points that relate to the information you're providing, while underscoring the benefits your product or service offers.

Letter 3–1

MATHEMATICA PRODUCTS GROUP, INC.
P.O. Box 2392
Colt's Run, New Jersey 00000

March 27, 19—

Mr. Joseph Morris
Arcwell Corporation
212 Kenyon Street
Arcwell, Michigan 00000

Dear Mr. Morris:

I'm sorry you were unable to join us at our recent RAMIS II seminar. Since you had indicated an interest in RAMIS II by registering for the seminar, this brief letter will help to acquaint you with our system.

RAMIS II is an innovative computer language with comprehensive capabilities for data base management, research, report preparation, and information retrieval.

Since its origin over ten years ago, RAMIS II has been expanded
and refined continually to include new features and benefits.
These improvements have helped simplify communications between
the user and the computer network. Research staff, laboratory
technicians, product designers, sales engineering personnel, et
al. are represented among the thousands of clients utilizing
RAMIS II.

The enclosed material provides an overview of the system. Because we're convinced that RAMIS II is of genuine significance to
the engineering professional, we hope you will give the enclosed
data your special consideration.

To further assist with your needs, a RAMIS II representative will
contact you within the next few days. Following your discussion
with Mr. William Thornton, if I can be of additional assistance,
I'd be delighted to hear from you.

Cordially,

Alternate Approaches

(a)   Thomas Kean, design engineer on our staff, told me of the problems
you've been having with your present data base system. Normally, we try
not to overlap the function of our marketing team, but in this case, I think
we might be of some help to you. Tom probably told you something about
our Ramis II system, and perhaps if this situation had occurred last year I
wouldn't be writing to you. Reason: improvements we've made in recent
months seem perfectly synchronized with the problems you've been having. So much so, in fact, that I believe Ramis II is now able to provide
solutions to each of the problems Tom described to me. I've enclosed a
detailed review of current design specs and invite you to compare these
features with your present system. I have also asked Walter Reston, one of
our sales engineers, to give you a call next week to see if we can be of
additional assistance.

(b)   There was a time when only our marketing people staffed the
product display exhibits at various association meetings. I'm glad we now
have at least one staff engineer in attendance; otherwise, I would not have
had the opportunity to chat with you in Memphis last week. I was most
interested in what you had to say about the software your firm is presently
using, and only regret that interruptions during our conversation prevented
a more detailed discussion. The material with this letter may help to fill in

some of the areas we didn't have a chance to cover. Naturally, as one of the design engineers on the RAMIS II system, I'm familiar with what it can do (and with what it should not be expected to do). This convinces me there is an opportunity here to make a test that could prove helpful in resolving some of the problems you described. Specifically, a test that could determine whether your firm would indeed save time and money by using RAMIS II.

(c)   Because of your obvious experience with a broad range of computer software, I'm going beyond a brief answer to your basic question. The additional material with this letter deals with systems related to data base management, report preparation, and information retrieval. Actually, all three of these important topics are blended in a package that could make a sizable contribution to your firm's profits. We can't be certain that RAMIS II would fill all your needs, but the potential is there . . . and it is one that would seem worth exploring carefully. As one of the engineers that contributed to the development of this system, my tendency is to dwell on the technical features. However, I'm aware that your concern is with two important aspects: Will it fill *your* needs in a cost-effective way and, indeed, contribute to your firm's profits? To resolve these related questions, I have asked one of our sales engineers, William Benton, to get in touch with you in the next few weeks. After you talk with Bill, if you have additional questions, please give me a call. I'll be glad to help in any way I can.

● *ANSWERING PRODUCT INQUIRIES*

General Guidelines

In a more leisurely era, before TV, letters of this type could depend to a certain extent on their entertainment value. Contrary to the approach used by Letter 3–2, colorful, extravagant claims and elegant introductions were often used to lure a prospective client down the primrose path to the ultimate message. Now, such approaches merely act as a barrier between the reader and the information you're trying to convey. If your message is not compelling in itself, it's unlikely a lot of froth and extravagant phrasing will make it so. The current rules are basic and can be stated briefly: Thank the writer. Be gracious and to the point. Enclose relevant literature or other requested data, but don't overwhelm the reader with excess illustrative material. Offer to send any additional information that may be needed.

Letter 3-2

WALDWIN ELECTRONICS INSTRUMENTS, INC.
412 Ames Drive
Dallas, Texas 00000

March 26, 19—

Mr. Allen Wright
Royal Data Corporation
14 Oak Street
Carswell, New Jersey 00000

Dear Mr. Allen:

We appreciate your interest in our products. It's a pleasure to
enclose the information you requested, along with some additional data we hope will be helpful.

Over the past decade, we've developed and produced hundreds of
thousands of semiconductor devices for the computer industry.
This experience has resulted in a wide range of technological
know-how and extensive mechanized production facilities that
could be of service to you, both now and in the years ahead.

I'll be glad to provide additional information or, if you wish,
you might contact our sales engineer who is responsible for customer service in your area. His name is Donald Wilson, and you
can reach him at (000) 000-0000.

Thanks again for your inquiry, Mr. Allen.

Cordially,

Alternate Approaches

(a) Some of our people are closer to the subject of user experience than I am, but I gladly enclose the technical data you requested. As a means of supplementing these data, two additional steps have been taken. First, the enclosed material also includes an overview of lab equipment we manufacture. As you will note, we cover a wide range, mostly geared to the needs of firms like yours. Second, and in direct response to your inquiry regarding user experience with the LX12 laser, I have asked one of our sales engineers,

Larry Thompson, to get in touch with you. I'm certain Larry can provide you with helpful information on this subject. Of course, whenever questions develop on technical matters, contact me directly. I'll be glad to help.

(b)　I suppose this might be considered a "good news/bad news" response to your inquiry. The bad news is that we discontinued production of the X1A model several months ago. The good news is that we did so because design changes led to the much improved Model X1B. The enclosed design specs and related brochure provides the essential information you need to make a comparison. You, of course, will be the ultimate judge as to whether the X1B will fill your needs. To help in this decision, I have asked our product manager, Henry Schafer, to give you a call next week. I do hope his assistance, coupled with the material that accompanies this letter, provides useful data for you. After your talk with Henry, if there's any additional way I can be of help, you have only to ask.

(c)　We do appreciate your interest in our products. Your request gives us an opportunity to share with you some of the latest innovations developed by our engineering staff. The data you requested are enclosed, although I regret we had no way of knowing which particular product or products you have a primary interest in. Since your letter indicates you are associated with the Benton Corporation as a mechanical engineer, I wonder if your principal interest is in our hydraulic presses. If so, I'd be glad to send detailed specs on our various models as well as any additional data on our products that might be of particular interest to you. We place a great deal of importance on letters such as yours, Mr. Jones, and I look forward to being of help to you in the future.

● *SUPPLYING ALTERNATIVE INFORMATION ON A PRODUCT*

General Guidelines

As reflected by Letter 3–3, the inability to fulfill a client's request presents a problem. There are two problems, actually: one for the person making the request, the other for the one responding to it. You have a better chance of overcoming or at least reducing any negative reaction by making the letter friendly and sincere while offering an alternate course of action. Briefly stated, the guidelines are to (1) tactfully explain why you cannot provide the data requested, (2) offer to help in another way, and (3) express appreciation for the individual's interest in your organization. In essence, project a genuine desire to help, even though you cannot offer the specific data indicated in the request. (Also, see Letter 3–15.)

Letter 3–3

ABRAHAM'S MANUFACTURING CORPORATION
112 Chestnut Street
Fort Davis, Maine 00000

April 16, 19—

Mr. Thomas Hilton
Hilton Machinery Corp.
1441 Elm Street
Arkland, Pennsylvania 00000

Dear Mr. Hilton:

Thanks very much for the letter of April 14 and your kind remarks
about equipment you purchased from us last month.

I'd be happy to rush a copy of the pamphlet you requested—if only
we had one. The unexpectedly heavy demand has completely exhausted our supply, and it will be a couple of weeks before I'll
be able to fill your request. In the meantime, the enclosed literature will acquaint you with two new products we expect to put
on the market in May. Because of the nature of your business and
the fact that these products represent an improvement over some
of the equipment you are presently using, you may want to investigate one or both of these products further. If so, I'd be glad
to send a couple of samples to you.

Once again, as soon as the reprinted pamphlets are available,
I'll rush a copy to you. In the meantime, thanks once again for
your letter, Mr. Hilton.

Sincerely,

Alternate Approaches

(a) Thanks very much for your letter of May 4 in which you ask for
information about our new series of workshop manuals. The fact that this
is indeed a "new" series prevents me from sending this information as
quickly as we'd like to. Because of various innovations introduced in this
collection, we want to make certain the copyright procedure is completed
before we distribute detailed information. In addition, there are a number

of last-minute improvements we want to make, and this will also require some additional time. All this will benefit both of us, because the data you eventually receive will more accurately reflect the unique and exceptionally practical value this collection offers. In the meantime, the enclosed publicity release will at least give you an overview of the series. Please be assured you will be among the first to receive a detailed brochure, and we'll make every effort to get this out to you in the next few weeks.

(b)   The information requested in your recent letter is not available to us. You can obtain this from the National Bureau of Standards, by contacting the U.S. Government Printing Office. I do not know if there is a fee involved, but I believe they have copies available. If they do not (which would surprise me) you might try the IEEE. In addition to these suggestions, and because of the nature of your company, I have enclosed data on some of our services which relate to products you manufacture. This is not meant as a marketing ploy, but is sent simply as an extension of the data you requested. If any of this material prompts additional questions, I'll be glad to help in any way I can.

(c)   Because of the rapidly increasing trend toward greater speed in telecommunications, we no longer manufacture a 300-baud modem. I can appreciate your reference to lower initial cost and suggest that you also evaluate long-term costs between 300 bps as compared with a modem capable of 1,200 bps. True, the initial cost is greater for 1,200 bps, but even if you only use the modem for a few hours a week, I believe you will eventually save money by going to 1,200 bps. You may want to contact the Benton Corporation for information on its 300-bps model. I have also enclosed detailed specifications on our MX1200A. Your inquiry is sincerely appreciated, Mr. Thornton. Let me know if you have additional questions. I'll be glad to help.

● *REQUESTING PROMPT PAYMENT FOR SERVICES RENDERED*

General Guidelines

Letter 3–4 represents one of the more delicate areas in the professional/client relationship. While this is true, particularly in the advanced stages of a credit problem, logic dictates that the more acute the problem, the less "delicate" the corrective action needs to be. Particularly with clients of long standing, however, a deft and friendly reminder may be all that's necessary. This is also a good way of telling the individual that although you are asking for payment of an overdue bill, you still regard him or her as a valuable client. In summary, mention the good relationship you have always enjoyed. If appropriate, offer to help arrange a more suitable payment schedule. Give the person an opportunity to save face. Express confidence in his or her integrity and sense of fair play.

Letter 3–4

APEX ENGINEERING CONSULTANTS, INC.
2200 Meade Lane
Orlando, Florida 00000

March 28, 19—

Mr. James Bentley
Treasurer
Cantwell Corporation
412 Phoebus Rd.
Nelson, Vermont 00000

Dear Mr. Bentley:

Our organizations have developed a mutually satisfactory relationship over a long period of time. Consequently, I hope you will understand that my motive is entirely friendly in writing this letter to you about something that has been giving us some concern.

Our records show that payments by your firm are reaching us increasingly late. We have always appreciated your firm's promptness in the past, and, both to you and to us, it is important that we continue receiving payments according to the terms we've agreed on.

If there is some serious problem confronting your firm, and if you will tell us about it, perhaps we can help. If, on the other hand, somebody in your office has been tardy in processing the appropriate payments, would you please see that our account is give closer attention in the future?

```
I hope you will accept our request in the spirit in which it is
intended and will work with us in your usual cooperative manner.
Please arrange to send a check for the past-due invoices as soon
as possible.

Thanks very much, Mr. Bentley.

Sincerely,
```

Alternate Phrases

(a)   Few people like to repeat themselves, and I regret the need to send a second reminder requesting payment for services we rendered almost two months ago. Because of the excellent relationship we've enjoyed in the past, I'm wondering if our earlier letter went astray. Regardless of the reason for the delay in payment, I do hope you will pause a moment and see that a check is issued today for the full amount. A second copy of the bill is enclosed, along with a prepaid return envelope. We value your account with us, Mr. Hall, and look forward to continuing our business relationship with you and your firm.

(b)   We have tried to be patient, as evidenced by the enclosed copies of our earlier correspondence. Despite the fact we have valued you in the past as an excellent customer with a fine credit record, every firm must collect its bills to stay in business. With genuine regret, we must now inform you that a time limit has been set for payment of your account in full, including service charges, for purchases made on March 12, April 4, and April 11. If payment is made before May 15, you can avoid having your account turned over to a collection agency.

(c)   The record indicates our firms have enjoyed a mutually beneficial relationship for some time now, one reason being the fact we have been able to communicate with each other. This letter is not only an expression of our appreciation for this relationship, but also an effort to avoid having that relationship adversely affected. The engineering services we provided for you during the week of May 12 were, I believe, performed promptly and efficiently. Payment was due no later than June 12, but as of today we still have not received your check. If payment has been delayed because of some difficulty you've encountered, please explain this to us so we can try to help in some way to resolve the matter. Good, long-standing business relationships are vitally important to both of us, Mr. Walker. We want to help ensure the continuation of ours just as much as we can. I'd appreciate your response to this in the very near future.

● *COMMUNICATING IMPROVED SERVICE PROCEDURES*

General Guidelines

Improvements in production and/or servicing procedures are not always communicated to the people most affected by them. Letter 3–5 represents one way to correct this problem. A good letter of this type can pave the way for those dealing with the individual who stands to benefit from these improvements. It's usually a good idea to keep the tone of the letter informal. When introducing someone who will follow up on your letter, briefly convey the qualifications and background of that person. Praise the individual; express confidence in his or her knowledge and ability. Emphasize the specific ways in which product and/or service improvement(s) will benefit the reader.

Letter 3–5

AMERICAN AVISCO CORPORATION
410 Abbott Way
Los Angeles, California 00000

March 28, 19—

Mr. Henry Thatcher, President
Mac Hughes Corporation
266 First Street
Maynard, Connecticut 00000

Dear Mr. Thatcher:

I'm glad to tell you that Mr. E. R. "Mike" Smith of the Technical Service Department will now work out of our Los Angeles office to give you prompt assistance with your packaging problems.

Mike, a specialist with Product Design and Packaging for over 20 years, has had broad experience in production, quality control, technical service, product design, and development. He is exceptionally well qualified and will be glad to respond to your questions and needs regarding packaging machinery, laminating, adhesives, and inks.

```
Now that Mike is located within your area, you will get even bet-
ter service from Avisco on the West Coast. However, whenever any-
thing comes up that requires my personal assistance, I hope you
will continue to contact me directly.

With best regards,
```

Alternate Phrases

(a)   You have not met her yet, but I feel certain you'll be pleased with the new architect we've assigned to the Wynard Mall project. Ms Helen Rogers is not "new" with our firm. She has come up through the ranks and is well prepared for this important assignment. Bill Wrigley, the supervising architect, is delighted to have her on his staff, and there's no doubt in my mind you will eventually share his high opinion of Helen's skill. When Helen reports for work on March 12, I hope you'll manage to spend some time with her, mainly to share your views and suggestions regarding the changes we agreed on last week. I have enclosed a copy of her bio data and requested her to give you a call next week to set up an appointment that's convenient for you.

(b)   It has become obvious that you are not getting the degree of support and cooperation that, as one of our most valued customers, you are entitled to. Consequently, you'll be pleased to know we've assigned Thomas Campbell to serve as project engineer for the Hamilton building. Tom has been carefully briefed on the problems you called me about last week. Naturally, we don't expect you to be "pleased" yet. But after you hear the proposals he has to make, I believe you will agree the solutions he has to offer are both ingenious and cost effective. I've enclosed a brief review of his excellent background. He is enthusiastically looking forward to this assignment, and as your secretary has probably told you by now, we've arranged for Tom to meet with you and your staff at 10 A.M. on Monday, June 12.

(c)   This is not meant to minimize your complaint. My purpose is to reinforce it, the point being, it was coincidental that you criticized the service you've been getting on the Bennington project. After seeing the preliminary plans, I arrived at much the same opinion you expressed, although I was probably more critical. No question about it, Roger, we need a specialist. And that's why we've retained Harvey Thatcher, who is probably one of the most experienced engineers available, and certainly one of the most skilled in the area of hydraulic valves. Harvey has examined all the original specifications and in the process spotted a variety of design errors. I have asked him to join us at our forthcoming meeting on June 4. He will be prepared at that time not only to identify the design flaws but to propose specific solutions, along with realistic estimates regarding the effect these solutions will have on time and cost factors.

● *DECLINING INTEREST IN PRODUCT OR SERVICE*

General Guidelines

A tactful letter explaining why you are not buying a company's product or service is a courtesy that's often overlooked. Letter 3–6 simplifies this type of communication and will help to keep the lines open to potential vendors. There may be practical benefits as well. Something in your response may open up another approach or idea that could be beneficial to both parties. The general rules are simple. Thank the vendor for the offer of merchandise or service. Explain why you have decided not to pursue the matter at this time. If appropriate, hold out the possibility of future business transactions.

Letter 3–6

BURLINGTON-PACIFIC CORPORATION
Equitable Building
Portland, Oregon 00000

March 26, 19—

Mr. Richard Abernathy
412 Thermo Drive
Keystone, Michigan 00000

Dear Mr. Abernathy:

Thank you for bringing your products to our attention. We buy a small amount of veneer now and also a variety of cabinet frames for our 13-inch and 19-inch television sets. Our use of veneer, however, is not extensive and probably will decrease rather than increase, due to recent connections made for the purchase of imported mahogany on a contractual basis.

No one, of course, knows just how fast our needs will develop, or how satisfactory our current supplier's performance will be. Consequently, it would be our pleasure to hear from you from time to time about the products you have available. However, at present we are in no position to make any definite commitments.

Sincerely,

Alternate Approaches

(a)   Many thanks for your interesting and informative presentation in my office last week. In discussing this with my staff, we discovered that our present inventory of lab equipment is sufficient to tide us over until the end of the year. We're hoping your special offers on such items as beakers and vinyl tubing will still be available when we decide to replenish our supplies in January. In the meantime, if you introduce additional products to your line, we'd be glad to hear about them. The material you left with me has been placed in our follow-up file, and I'll be in touch with you again shortly after the turn of the year. Thanks again for your recent visit.

(b)   The catalog and brochures you recently sent were very interesting. We know that yours is a first-rate company manufacturing first-rate products. Normally, we'd be happy to deal with you. The fact is, however, we have been doing business with another machine parts supply house for many years and are quite satisfied with their products and maintenance service. We are not, therefore, planning any change in the near future. When and if a situation arises that suggests we should discuss the matter further, you may be certain I'll get in touch with you. With this possibility in mind (even though it appears somewhat remote at this point), we'll retain your catalog in our files. Thanks once again for sending it to us.

(c)   It is not common knowledge yet, but we are eliminating two of our product lines. One of them contributed to our former need of piezo elements, which we purchased for use in our security alarms. While this action will not completely eliminate the need for your products, it will cut down on the amount we use. Consequently, we have a supply on hand now that will be sufficient for the next three months. Your recent inquiry prompts me to be frank with you about this, because I don't want you to think we were dissatisfied with your product or the service we received. As you know, we're constantly upgrading our product line, and these changes could create a need for other products you might provide. If so, I'll be glad to contact you for detailed price quotes. In the meantime, you'll at least know the answer to the question raised in your recent letter.

● *STIPULATING CHANGES IN PRODUCTION PROCEDURES*

General Guidelines

The need for communications such as Memo 3–7 often result from a complaint, either from within the organization or from a client. In this instance, both parties are involved. To ensure a proper response, the memo or letter should be brief and crystal clear. The sample memo and alternate approaches are short, direct, positive, and insistent, but still polite. In situations where the request or instruction is being repeated, a copy of the earlier request may be enclosed. In addition to indicating specifically what is wrong, your viewpoint should be expressed in a reasonable manner. Any reference to adjustments, corrective steps, or alternate courses of action should be equally specific.

Memo 3–7

March 28, 19—

TO: Mark Stevens

FR: Alan Jones

Mark,

An important client, Acme Manufacturing, is still complaining about our aqua blue paint fading in less than a year. We have been over this problem several times before, and until we can create a new formula, you will simply have to use the more expensive Z pigment. It was my understanding that you were already using it.

```
Please let me know when you reintroduced the Q pigment and why.
I would like you to change to the Z pigment immediately.

Alan
```

Alternate Approaches

(a)   In the past three months there has been a substantial increase (34 percent to be exact) in our expenditures for rollers purchased rather than manufactured. This has had a negative effect on our profit forecasting procedures, contributing to significant variances in the budget we all agreed on last June. In the future, before a commitment to purchase rollers is made, I want to review the cost, price, and profit relationships before any final decision is made. It appears that in some cases we don't even reach the break-even point and are actually losing money. My review will continue until it becomes certain that these outside purchases are contributing to profits. If sufficient evidence to confirm this is not generated within one month, instructions will be issued to ban all these outside purchases.

(b)   The facts will not allow me to accept your answers regarding the attached manufacturing complaints. Your statement that the valves are not likely to leak does not square with the obvious fact they have, indeed, leaked on the occasions indicated. As you will note, the Ames Corporation is asking for a credit of $849.00. Please provide a more detailed and effective response to this matter: Which crews and/or individuals are responsible, what is being done to prevent future complaints, and what procedural changes are you implementing to make certain the necessary changes become standard policy? Please return the enclosed complaints with your answers no later than April 20.

(c)   We are having problems with the air hammers produced by your department. Before the situation becomes widespread and even more costly, I would like you to take corrective action. In essence, dust is being taken into the lubricating system, causing the emergency cutoff switch to be activated. After considerable experimentation, we discovered the hammer can be operated, but only if the speed is substantially reduced. This simply is not acceptable. There's no way we could meet our production requirements under these conditions. Would you please, just as quickly as possible, investigate and correct this problem. Obviously, an effective filtering device must be developed to prevent dust from entering the system. I would appreciate it if you'd send me a written report, no later than May 15, on corrective action you've taken.

● *RESPONDING TO SERVICE COMPLAINTS*

General Guidelines

Letters involving complaints from colleagues or clients, such as Letter 3–8, often represent opportunities in disguise. No one enjoys being on the defensive, so any apology should be quickly followed by an upbeat, positive assertion as to the corrective action being taken. The wording may depend to some degree on company policy. Regardless of the circumstances, your response can determine whether a particular client is lost to you and your firm or, in fact, becomes an even stronger and more valuable contact in the future. The engineering staff can be just as important as a customer relations department in maintaining goodwill toward the company, and in some instances more so. This is especially true in matters requiring technical expertise.

Letter 3–8

MICRO ELECTRONICS CORPORATION
616 Ames Drive
Waycross, Georgia 00000

March 28, 19—

Mr. Matthew Wilson, Chief Engineer
General Controls Corporation
144 Wabash Rd.
Chicago, Illinois 00000

Dear Mr. Wilson:

This will acknowledge receipt of your letter concerning problems experienced with the X-400A computer. We regret the circumstances which prompted you to write. I'm glad your letter was

directed to the Engineering Department, because we have a vital
interest in making certain that clients are satisfied with the
products we produce.

To make certain the problem receives prompt attention, I have
forwarded a copy of your letter to the dealer and also to our
branch office nearest you. One of our technicians who is a specialist on this particular computer will contact you shortly. He
will also provide you with the name of the individual at the
dealership who is best equipped to answer your other questions.
First, however, we want to make certain the computer system
quickly receives any repairs that it needs.

The dealership will also review this matter with our district office, and they will advise us when all necessary and appropriate
action has been completed.

If these steps do not resolve the situation to your complete satisfaction, Mr. Wilson, please let me know. We appreciate your
business and want to make certain our cordial relationship continues in the future.

Sincerely,

Thomas Ajalat, Manager
Engineering Department

Alternate Approaches

(a)   You are correct. We made an error when itemizing the perforated
boards you purchased on May 14. This resulted from an earlier decision
to discontinue production of the 11 1/2″ × 12″ size. When you brought this
to our attention, we referred back to your original order and discovered
you had correctly indicated the new size of 10″ × 14″. I realize that giving
the reason for an error doesn't correct it, but hope you will accept our
apology and allow us to try harder to serve you in the future. You may be
certain we will do just that. Thank you for bringing this to my attention,
Ms Rogers. A corrected invoice will be prepared and mailed to you in the
near future.

(b)   The old saying has it that there's safety in numbers, but your letter is a good reminder that old sayings are not always true. The fact is, a considerable number of people reviewed the specifications for the enclosure design (RX200) we sent you on March 4. Not one of them picked up the error referred to in your recent letter. Your thoughtful comments are sincerely appreciated, Mr. Douglas. I'm not sure I would have been as tolerant if the circumstances had been reversed. Contrary to the impression you may have formed, experience indicates the RX200 does indeed deliver outstanding performance, and we would welcome your further consideration of the purchase you referred to. Thank you once again, and I have enclosed an additional (and corrected) set of specifications.

(c)   We sincerely regret that the DC power connectors you purchased did not meet your expectations, even though it appears the problem was not caused by our Production Department. We carefully control the quality of our products throughout the manufacturing process, and following this, each unit is double-checked by the Quality Control Department. Sometimes during the handling and placement in retail establishments the pin connectors on our DC power units are damaged, and this is a factor over which we have little or no direct control. Because we feel we cannot accept responsibility for damage after the product leaves our plant, we suggest you check with the supplier you purchased the units from. I hope this will result in a satisfactory resolution of the problem because we do appreciate your interest in our products.

● *PROVIDING ASSISTANCE TO CORRECT PRODUCT DAMAGE*

General Guidelines

Letter 3–9 deals with a situation in which one must walk a verbal tightrope. Product damage or service difficulties are not always the fault of the provider. Yet the provider would do well to avoid accusations that pin the blame on the client, at least until the facts have clearly determined that is the case. In this instance, the writer states her sincere desire to arrive at a fair and equitable solution, while expressing regret for any inconvenience caused the client. Letters of this type should be straightforward and courteous if they are to soothe ruffled feelings. Equally important, they should convey a genuine desire to resolve the problem quickly.

Letter 3–9

ELECTRONICS COMPONENTS, INC.
406 James Boulevard
St. Louis, Missouri 00000

October 16, 19—

Mr. Martin Guthrie
Apex Corporation
444 Carver Lane
Glaston, Virginia 00000

Dear Mr. Guthrie:

We are sending our chief technician, James Hatton, to look at the damaged circuits you received on September 27. An appointment has been arranged with your assistant. Mr. Hatton has been asked to fully resolve the question of whether a manufacturing error or shipping damage caused the problem you described. A determination will then be made as to whether a credit is issued or the entire order is replaced with new circuits. This aspect will also be discussed with you when Jim arrives at your office at 2:00 P.M. on Tuesday afternoon, October 27.

I regret any inconvenience this has caused you, Mr. Guthrie, but James Hatton will set things straight. We value your business and friendship.

Sincerely,

Alternate Approaches

(a) I hope it will be of some consolation to you to know we are as anxious as you are to uncover the source of the problem you described. We are as concerned about the way customers view our products as we are about the products themselves. In direct response to your letter, I have asked our service manager, Martin Baker, to investigate this entire matter carefully. He is most familiar with the operation of our digital voltmeters and is well qualified to make a fair judgment on this issue. While shipping damage cannot be ruled out, you can be sure that if the fault lies with the units themselves, we will replace each one. If, indeed, the damage occurred while en route to your firm, we will join with you in seeking appropriate action from the freight carrier. Mr. Baker will call you shortly to set up a convenient appointment time.

(b)   The Hex key sets you returned on May 12 have been received, and this has enabled us to identify the source of the problem you referred to. It was caused by the inadvertent use of an incorrect catalog number. If you will refer to page 211, you'll notice there are two products listed in the upper right-hand corner of that page. One is an 18-piece Hex key set with the catalog number 842–111, and your order indicates you wanted 25 of the sets that carried this catalog number. However the other is the more versatile set (catalog number 842–112) containing two Phillips screwdrivers, three Hex keys, and the torque bar referred to in your note. I have enclosed a revised invoice showing the differential in cost. Because of the inconvenience caused you, we will not wait for your check and the new sets will be shipped to you today. Please remit within ten days after you receive them.

(c)   The damaged amplifiers you received resulted from a compounded error that falls within the "one in a million" category. Instead of sending you the new MR400 models you ordered, our Shipping Department somehow managed to send you units that had been returned to us because of damage incurred during shipment. I cannot recall this ever happening before and will most assuredly take steps to make certain it doesn't happen again. Because of the regrettable delay in filling your order properly, I have arranged to send the replacement units via Air Express. We will, of course, absorb these shipping charges. Thank you for bringing this to our attention, Mr. Wainwright.

● *SUGGESTING CHANGES FOR INCREASED PROFITABILITY*

General Guidelines

The somewhat cautious, deferential tone used in Letter 3–10 is a good one to take when you're not certain how your suggestions will be received. The danger in writing such a letter, even though your suggestions may have been requested, is that you may sound too critical, conceited, or pedantic. There may even be a possibility that the individual asking for suggestions really doesn't want them. In this case, the easy way out is to gear your ideas to factors over which the reader has little control—and to say so. This is hardly the most constructive approach, and one that could easily boomerang. If the reader is important to you, then he or she deserves a determined effort on your part to be helpful.

Letter 3–10

OMEGA COMPUTER SYSTEMS, INC.
1212 Fairchild Avenue
Hillsdale, Illinois 00000

April 25, 19—

Mr. Henry Smith
Engineering Consultants, Inc.
1046 Rogers Street
Suburbanville, Georgia 00000

Dear Henry:

I was impressed with your plans for the consulting operation and the variety of services you intend to perform, but I think you are making a mistake to hang onto that location. Old rules of thumb like "it takes two years to get a business under way" can lead to disaster if you cling to them too slavishly. You yourself said the location is not the best, so why not change it, and change it now? Would it not be worthwhile to take some time out and find a really good spot?

Chain stores may not be the best example, but I think they prove that the "two years" stuff is a lot of baloney. Charles Morse, who I think you met a while ago, said that he made a personal commitment when he went into the consultant business. Specifically, if he wasn't turning a modest profit in six months, he'd get out. That was three years ago, and he is the first one to admit that the location he picked was one of the deciding factors that enabled him to turn the profit corner in less than six months. Of course he spent a good deal of time in making marketing studies, particularly with respect to office location, but there's no reason why you could not do more of this yourself.

I hope you don't think I'm being too forward in urging you to give greater consideration to this aspect. You have so much

ability and the potential is so great that I hate to see you trying to progress under less than optimum conditions.

Let me know what you decide to do. I'm genuinely interested.

With best regards,

Alternate Approaches

(a) While spending a good deal of time in your production department recently, I noticed you are still using the LIFO system of inventory record keeping. You doubtless adopted this a while ago, possibly during a period of rising prices. I can't help but feel, however, that in a period of intensified competition and declining prices, you might find that the more conservative FIFO system offers a number of advantages. I would be glad to point out the good features of this system if you're interested.

(b) I am honored by your request for my opinion on the conveyor belt operation. Actually, I do have a suggestion, but it is somewhat peripheral to the conveyor belt layout. With respect to the latter I believe you've done a fine job in designing this for maximum efficiency. My thought deals mainly with the present location of the storage bins and their proximity to the main transfer points on the belt operation. I believe this represents a problem that could seriously affect operation of the entire system. If, on the surface, you agree this is something that warrants more detailed consideration, I'd be glad to meet with you next week. It would also be a good idea to have your plant manager and systems supervisor at this meeting. Provided you want to discuss this further, give me a call, and we'll set a convenient time and date.

(c) It was coincidental that strong emphasis was placed on product design during our recent staff meeting. I had discussed this the day before with one of the marketing people (Tom Keating). The more I think about it, the more convinced I become that the carton design we're using for the emitter/detector set is misleading. Aside from the design itself, the printed material should really have been checked and double-checked by someone in the Engineering Department. There is, for example, no mention of the fact this set can be used to send analog *or* digital signals through fiber optic cable. Nor is any reference made to the length of the fiber optical cable included with the set. Purely from an aesthetic standpoint, the colors are drab and do not even complement each other. I hope you agree we should bring this up with the appropriate people at the earliest possible date. Please tell me in the next day or so how you'd like to proceed on this.

● *EXPLAINING REASONS FOR DELAYED*
   PRODUCT DELIVERY

General Guidelines

Letter 3–11 tackles a common problem, and one that seldom permits an explanation that is totally satisfactory to the client. The primary need in situations of this type is corrective action, not explanations. Yet explaining the reason for a delay or an incorrect action can be an informative courtesy to the client, particularly if it is combined with assurance the problem has been, or is in the process, of being corrected. No need to dwell at length on the cause. State the facts briefly and clearly. Follow this with specific reference to the alternative or corrective action being taken to overcome the problem.

Letter 3–11

ATWATER COMPUTER CORPORATION
412 Ascot Drive
Milwaukee, Wisconsin 00000

April 24, 19—

Mr. Joseph Anderson, Supt.
Materials Department
Modern Methods Services, Inc.
Big City, Kansas 00000

Dear Mr. Anderson:

The local transportation problems that have been in the news these last several weeks have played havoc with our production schedules. Some components of the D14JL computers we were sup-posed to install for you by April 20 have not been delivered to us even at this late date. Consequently, it will be impossible for us to deliver on time—in fact, under the circumstances, I don't want to mislead you by estimating even an approximate date for delivery under the present conditions. We are greatly

```
distressed at the way things have gone, and sincerely hope that
no great dislocation is caused by this event. You may be certain
I'll keep you posted on any developments, but please be assured,
we'll get these components to you as quickly as possible.

Sincerely,
```

Alternate Approaches

(a)   The special dies required for Project X110 were delivered to us several days late by the supplier, and I regret this as much as you do. It will require us to go on an overtime schedule, and we will have to absorb this cost. Even with this additional effort on our part it is possible that the X110s you ordered will not be completed by June 10. There is no way we can presently estimate how much extra time we'll need. Certainly we will exert every effort to make up for lost time, and my guess is you'll have the entire shipment no later than June 15. I'm waiting for a report from our production manager and can then give you a more precise shipment date.

(b)   We both know of firms that think nothing of agreeing to a deadline date for delivery and then blithely let the date pass without alerting the customer to this in advance. They rely on stalling the customer and on vague, flimsy excuses to cope with the situation. By doing this, they often throw the customer into an emergency situation and irreparably damage their own relationship with the customer. We don't operate this way. You placed your order #22462 on May 15, we promised shipment on June 15, and it is now May 22. It now appears more likely that actual shipment of the complete order will be made June 18. Reason: our supply of the high-output infrared LEDs has been reduced by a work stoppage at the supplier's plant. I have been told the matter has been resolved and full production will be resumed tomorrow. If this is true, and I have been assured that it is, we will send at least half of the shipment on June 15, and the remainder on June 18.

(c)   A copy of your original order is enclosed. Please note the third item, #44214. We included production of this in our estimate which was submitted to you on March 12. Now, we understand from your Purchasing Department you would like to omit this item from the original order. We are agreeable to this, but I just wanted to notify you that the action will cause a delay in delivery of the remaining items. The reason is that retooling and setup problems caused by this omission will take additional time. Under the circumstances, there was no way we could have taken this into consideration when the original time and cost estimate was submitted to you. If, indeed, this causes a delay, it should not be more than a few days. I'll be in touch with you again just as soon as we're able to give you an accurate delivery date.

● *EMPHASIZING PRODUCT/SERVICE ADVANTAGES*

General Guidelines

In effect, Letter 3–12 sets the stage for reconsideration of an earlier decision, by reemphasizing the product's advantages. There is a wide range of situations that could be covered by adaptations of this letter, enabling you to convince or reassure the client on the value of the product or service you're describing. As with most other letters in this chapter (and throughout the book for that matter) brevity plays an important part. Many points are never made, and many sales are lost because the one trying to make them doesn't know when to end his or her presentation. Compelling facts, concisely stated, will immeasurably strengthen your case. Do not plead, and do not try to intimidate the reader.

Letter 3–12

ARDSLEY MANUFACTURING CORPORATION
840 Sands Road
Cleveland, Ohio 00000

February 4, 19—

Mr. John Jones
Purchasing Agent
Eastern Manufacturing, Inc.
Hillsdale, New Jersey 00000

Dear Mr. Jones:

After talking to you yesterday, it occurred to me that I had not placed sufficient emphasis on the actual money savings achieved

with a copier that can use any kind of paper and does not require
special coated stock. Our studies confirm that the extra cost of
special paper actually amounts, in less than a year, to a sum
equal to the purchase price of our Magic Copy Machine.

I have enclosed a detailed analysis showing comparative costs of
paper used by various machines, so you can go over them and check
them against your own experience. When you look at these figures,
also not that the material itself is printed on an ordinary letterhead of medium quality—uncoated. This is an example of routine work done on the Magic Copy, yet it's so clear and clean it
could be used for a mailing piece—far clearer, as a matter of
fact, than most copies on special coated stock coming from other
machines.

I would have underscored this cost-saving factor during our
meeting, but wanted the additional information provided by the
detailed analysis enclosed. Now that we have these hard figures,
I'm hoping you'll review your decision. If you would like a
demonstration on your own letterhead or on any paper at all,
we'll be happy to deliver a brand-new machine to your place
at any time convenient to you. You can reach me at 000-0000,
Ext. 00.

With best regards,

Don Hall

Encl.

Alternate Phrases

(a)   Your request for the larger size SPDT relays is understandable because you used this size on the Model 400. Yet there appears to be a way
you can save some space on the chassis, while saving enough money to more
than offset the cost of larger switches we discussed. Specifically, there is a
mini SPDT relay available in the following dimensions: $9/16'' \times 25/32'' \times 5/8''$.
In lots of 5,000, the cost would be $2.49 per unit, which would save you
over $4,000. The mini version is also rated at 3 amps at 125 V AC. Thus, the
specs are all right, the space saved would allow more room for the elec-
trolytic capacitors, and you would then be able to use the larger switches.
This sounds like a worthwhile change to me. Shall we go ahead with purchase of the mini version?

(b)   As pointed out during my presentation on Tuesday, robotics represents a "mysterious" subject only to those who have had little experience in this field. It is true that the initial cost for the installation is high. But you know what your costs are now using the manual labor approach, and fortunately, you have a clear idea as to how you would apply the Armatrix 1210. We also estimated the break-even point following our discussion, and I have enclosed a cost analysis that covers this aspect in detail. When you take into consideration the substantial savings in plant costs (not to mention the hidden savings in terms of sick leave, absenteeism, benefits, etc.), we're talking about a solid and sustained contribution to profits. And this contribution would become evident within six months following the installation.

(c)   It was a pleasure meeting with you yesterday morning. It was a learning experience for me, and I hope you also found some of the data we provided informative. As we discussed, digital radiology provides various advantages over conventional radiographic methods. There are many, but just a few would include improved diagnosis, reduced radiation exposure to patients, and lower operating costs. We can also provide you with methods for converting film images into digital images and advise you on specification and operation of laser scanners and other film scanners. An additional feature to consider would be the matter of analog-to-digital conversion techniques. In all, the benefits add up to a compelling rationale for following through on the changes we discussed. I look forward to being of additional service to you. If you need additional information before discussing this with your board next week, just give me a call.

● *OVERCOMING OPERATIONAL PROBLEMS*

General Guidelines

Letter 3–13 could be applied to difficulties arising from production, service, design, and/or delivery problems. By acting quickly, and making your action part of the written record, you stand a chance of bringing corrective action before the situation becomes more costly. Also, in some cases the supervisor, employee, or vendor may not even know there's a problem until you alert him or her to it. Be specific about the nature of the problem and what you want done about it. Be objective in describing the situation. Don't make the proverbial mountain out of a molehill. If you have sufficient knowledge, don't hesitate to make suggestions for correcting the problem. An acerbic approach is counterproductive. State the facts concisely, clearly, and courteously.

Letter 3–13

WRIGLEY MANUFACTURING CORPORATION
480 Wainwright Lane
Cauley, California 00000

May 15, 19—

Mr. Walter Jones
812 Tasley Street
Peoria, Illinois 00000

Dear Mr. Jones:

Because of a failure to agree on terms, the trucking firm that handles our deliveries has called a strike which may affect our operations. Despite this, our company intends to do everything possible to maintain normal shipping operations from our warehouse. We are now making alternate arrangements to ship the solid-state equipment you ordered on May 10.

You can be certain our first priority will be to minimize any disruptions that might inconvenience our customers.

Your understanding during this period is appreciated.

Cordially,

Alternate Approaches

(a)   The reputation of an engineering staff can be one of its most important assets. I prize ours, and feel certain you do the same. That's why we need to work together in quickly resolving the production difficulties discussed at yesterday's meeting. The heart of the problem, as I see it (and as supported by the facts), is John Kiley. Chronic absenteeism appears to be only one element in his poor performance. Another is his lack of commitment to staff objectives. Your version of the situation convinces me you've made a reasonable and concentrated effort to change John's attitude and improve his performance. Obviously, he has not cooperated. I recommend you immediately assign him to probationary status. If his performance is such that he does not meet his production quota within the next month, my suggestion is that he be released. Please let me know by Monday whether you agree with this, and if you do not, what do you recommend?

(b) The defects brought to my attention by management are detailed in the attached data. I have also included the design engineer's report, which offers his explanation of these defects. Now, it's your turn. And logically, you are the one who should be most aware of these problems. The quality control manager should not only be concerned with the quality of our products, he should also be concerned about anything that exerts an adverse effect on that quality. Frankly, John, I don't think the present situation reflects that awareness. I hope you will consider our meeting yesterday as a prelude to action—your action. Please let me have your recommendations no later than Monday of next week, and be prepared to discuss them with me in my office on Tuesday at 10 A.M.

(c) This letter alerts you to a serious production problem affecting delivery of components for the instruments we're providing to the Air Force. A delay in delivery of these components will seriously compromise final delivery of the instruments and could adversely affect our future relationship with this vitally important client. The Air Force Contract Management District in New York is being informed of this possible delay by a copy of this letter. We therefore urge you to take all necessary steps to assure the Air Force and ourselves that delivery will be made as contractually required. Please assure us by return mail that you are giving this problem your personal and urgent attention.

● *EXPRESSING APPRECIATION FOR PRODUCT SUPPORT*

General Guidelines

Letter 3–14 supports the contention there are few goodwill builders as basic as the letter of thanks for a favor or thoughtful act. Yet busy engineers as well as other professionals often find themselves so rushed that the opportunity slips by and the letter or memo doesn't get written. Consider for a moment the probable ratio between letters of criticism and letters of appreciation. Human nature being what it is, the ratio undoubtedly tilts toward the critical communication. This makes the thank-you note exceptional and increases the likelihood it will make a lasting impression on the recipient. After making it clear what you are thanking the reader for, express your gratitude in a sincere way without being effusive. Whenever possible, let the reader know the results of his or her thoughtful act.

Letter 3–14

THORSEN ELECTRONIC DEVICES
Elmwood Park
Boston, Massachusetts 00000

March 3, 19—

Mr. Maxwell Benton
A. G. Roberts Co., Inc.
Burlington, Vermont 00000

Dear Mr. Benton:

I want to thank you and A. G. Roberts Co., Inc., for your gracious hospitality in allowing me to make a presentation of our new color graphics equipment to your Engineering Department.

The response from your staff was most gratifying. Moreover, the questions asked after the presentation were challenging as well as meaningful. I couldn't have asked for a better audience.

If you or any of your staff have additional questions about our color graphics system, it would be a pleasure to provide additional information.

Sincerely,

Lois Maxwell

Alternate Phrases

(a)   We're especially proud to have people like Andy Wilson on our engineering staff and very happy that people like you take the time and trouble to write us about your experience with our organization. Andy was a key player on the team that developed the Model 100, and it was understandable you singled him out when writing your letter. The fact you are one of the clients we value most made your letter even more gratifying. Rest assured we'll continue to do everything in our power to ensure your continuing satisfaction with out products. Thank you very much for your thoughtfulness.

(b)   I want to write you now to tell you how grateful we are for your exceptional performance on the Product Improvement Committee. No group ever had a more capable or more dedicated member than you, Tom. In the four years I've served as chairman, I've watched this Committee regain its strength, its confidence, and its purpose. There is no doubt in my mind that you are the one who deserves a substantial share of the credit for this. Your most recent contribution to the redesign of the 840 scanner was especially impressive. Everyone connected with this product (most of all, our customers) owe you a debt of gratitude. Sincere thanks, Tom. We look forward to having you with us again during the coming year.

(c)   Your sketches for the Oak Mill development caused us to take a much closer look at the problems described by the building superintendent. I can't tell you how much we appreciate the obvious thought and effort you put into this. I can, however, send this note of thanks—and gladly do so. You not only pinpointed solutions to the major problems, but the way you did it clearly reflected the solid support you've given this project all along. It was just another reminder of how invaluable your support is to us. As a more tangible expression of our appreciation, I hope you'll plan to be our guest at a staff luncheon on Tuesday, May 12.

● *SUGGESTING WAYS TO RESOLVE PRODUCT/SERVICE REQUESTS*

General Guidelines

As illustrated by Letter 3–15, goodwill can also be gained by suggesting other approaches when you are unable to fulfill a client's particular request. This letter and the alternate approaches supplement an earlier section in this chapter dealing with the supply of alternate data. Refusing a request for information, for whatever the reason, should be done politely. The request may cone at an inconvenient time or may even appear frivolous or pointless to you, but if the requester has gone to the trouble to single out one company or person, it is obvious the request is important to the person making it. Respect for, and a show of interest in, that person can go a long way toward building goodwill for you and your organization. Offering practical assistance clearly reflects your desire to help. In addition, it will help you ensure that individual's continued interest and/or support of your products.

Letter 3–15

DELACOURT CORPORATION
515 Piermont Avenue
Albany, New York 00000

March 3, 19—

Mr. Dennis Woodward
Tech-Design, Inc.
22 House Street
Milwaukee, Wisconsin 00000

Dear Mr. Woodward:

Thank you for writing to us with a request for copies of drawings for the Number 5543 Ward Press. I am sorry we no longer have these drawings.

While there is no assurance it would be productive, you might try contacting Mr. Leonard Brown, managing director, Society of Technical Archives, 000-15th Street, Washington, D.C. 00000. His telephone number is (000) 000-0000. This society at one time had a wide range of research data that contained illustrations of machinery manufactured in the 1800s. I hope this suggestion will be of some help to you, Mr. Woodward.

Sincerely,

Alternate Approaches

(a) We're enclosing a tally of the monthly services we provided during March 19—. Unfortunately, we'll be unable to provide the details on monthly totals you requested because our accounting system does not retain the itemized entries you referred to beyond the current month. Nevertheless, the tally enclosed, because of the extra details we've added, may be of help to you. I sincerely hope so.

(b)   Your request for 25 copies of our manual on speechcoding has been received, and we do appreciate your writing to us about this. If we still had copies available I'd be glad to rush them out to you, but unfortunately our supply is exhausted and it's unlikely we'll order a reprint in the near future. There are two alternatives you may want to consider. First, you might write to the Superintendent of Documents in Washington, D.C., and ask for a list of publications they have available on this topic. Next, Prentice Hall publishes an excellent book titled *Practical Approaches to Speechcoding* by Dr. Panos E. Papamichalis. Dr. Papamichalis is on the faculty at Rice University and also serves as a member of the technical staff with Texas Instruments. The publisher's phone number is (201) 767–9620. I hope this will be of some help to you Mr. Jones.

(c)   Few things bother me more than having to tell a valued client we can't fill his request for information. I'm still going to try to help you, but the specifications on our Model 44 oscilloscope would be of little assistance at this point. We discontinued production of this model over three months ago. I have, however, enclosed specs on the Model 44A and a brochure that points out the advanced features offered by this model. In addition, the Heathkit Corporation in Benton Harbor, MI, is a firm you may want to contact. From the information contained in your letter, it appears it may have a scope you would want to consider. We do not produce oscilloscopes in kit form, but it does. I hope this will be of help, Mr. Rogers.

● *DECLINING OFFER OF ASSISTANCE ON NEW PRODUCT*

General Guidelines

Letters of acceptance or refusal, as illustrated by Letter 3–16, need to be crystal clear. Some require especially careful wording to avoid possible legal entanglements. Even if this aspect is not a matter of concern, there is still the need to avoid ambiguity or a trace of annoyance in the tone and content of your message. In any situation where you suspect there may be a legal angle to be concerned about, you are advised to get expert legal counsel. Tact is particularly important when declining an offer of assistance from either a colleague or client. Wherever appropriate, leave the door open in the event conditions change and you decide to seek the assistance that was offered.

Letter 3–16

MICRO ELECTRONICS CORPORATION
616 Ames Drive
Waycross, Georgia 00000

April 25, 19—

Mr. Richard Roe
Kentwell Manufacturing
2027 Myron Road
Janesville, Illinois 00000

Dear Richard:

It was good of you to offer assistance, and I appreciate it. However, at this point I must first work out things within our own organization. The staff is helping me map out plans now, and I will let you know what develops. You are well aware of the aggravation and problems with the introduction of a new product, and this makes me appreciate your offer even more.

Naturally, if we're not able to come up with a viable plan by using our own resources, I may contact you again about this. In essence, I am asking for a rain check.

Thanks again, Richard.

With best regards,

Alternate Phrases

(a)   Your offer to assist in the installation of our research data bank is appreciated, but the picture is not quite as bleak as I may have inadvertently painted it. Recent additions to our staff have brought additional skills in this area, and it now appears we may meet the deadline after all. We won't know this for sure until the end of next week, but I didn't want to wait until then before thanking you for your note. If it's all right with you, I'll wait until Thursday and then give you a definite answer. It's nice to know we can call on you if it really becomes necessary to do so. Thanks very much, Dick.

(b)   Thanks for telling me about the components you could make available for the updated version of Model 24x. Initially, I was tempted to take advantage of your offer, but the staff engineer on this project convinced me we can hold off on this, at least for the next few months, the reason being that we have more stock on hand than I had thought, which came as a pleasant surprise when I was told about it. As a result, it now appears we can save some time by using the supply available, without waiting for delivery of parts from your warehouse. Your offer was appreciated, Al. If I can return the favor at some point in the future, just let me know.

(c)   Providing service on the X100 involves so many governmental and legal requirements I'm afraid we'll be unable to accept your offer. I have no reason to feel your organization could not meet the necessary requirements. On the contrary, my impression is that this would represent little or no problem for you. The difficulty here lies with the time factor. And by the time we waded through all the necessary paperwork, we'd be past the deadline for completion of the work. So, I'm afraid we'll have to move ahead on this with our existing staff. On another, but somewhat related matter, we may have need for your consulting services with the conveyor system in Warehouse #2. We do no government contract work in this building, and the plant engineer may want to discuss needed improvements on this system with you. I have a meeting scheduled with him for May 2 to review the problems involved. If he is amenable, I'll ask him to call you and set up an appointment so the two of you can go into this in more detail. In the meantime, thanks for your offer on the X100.

● *CAPITALIZING ON COLLEAGUE'S DESIRE TO HELP*

General Guidelines

Memo 3–17 capitalizes on an offer by accepting it, but that is not the only way to benefit from this type of situation. Even if you choose not to avail yourself of the services offered, there are still reasons for declining in a courteous, graceful way, as illustrated by the preceding discussion in this chapter. When your answer is yes, the acceptance should reflect your gratitude to the person volunteering to help, and where necessary, indicate the parameters where the help should begin, and end. In some instances, you may also want to ask the individual for a reply to your letter, just to make certain there is no misunderstanding.

Memo 3–17

April 25, 19—

To: Jennifer Jones
      Records Department

FR: Mike Engelhart
      Research Lab

I appreciate your offer to sort out and make copies of all the lab inventory records for this year to date. I would be most grateful if you would proceed on this and send copies to me by the end of this month. On second thought, it would be better if you first gave me an estimate of the time involved. Then, if the Management Committee accepts this estimate and realizes it will delay completion of the assignment Ed Jones gave you, you can go ahead with the assurance that everyone is in agreement.

We're working against a deadline in providing this information, and without your help, we would not be able to complete the study on time.

Thanks very much, Jennifer.

Mike

Alternate Approaches

(a)   It was great of you to offer to pick up and deliver all that lab equipment to Atlanta for me. I know I would not have been able to to it because of the staff meeting on Tuesday, and the client would have had every right to be annoyed. Thus, you not only did me a big favor, but you probably saved us the goodwill of the client too. Thanks again, Tom. I hope you were not just being tactful when you said you had to go to Atlanta anyway. And I hope I can do something to repay you in some measure one of these days. When such a situation arises, don't hesitate to ask.

(b)   Your offer is appreciated. Frankly, I don't like to put you to the trouble of sorting out all the research data we discussed, but you can be sure I'll be grateful for your help. The icing on the cake was your solution to the primary problem, and I agree that this will help simplify the sorting process. In terms of the time factor, let's approach it this way: if you find it's going to take longer than three days, let me know and I'll ask Bill to delay completion of the systems analysis he's working on and help you. He may not be delighted with the request, but I'm sure he'll do it. As you know, we're working against a deadline here, and there would not even have been a slim chance of our coming in on schedule without your help. Thank you, Eleanor. Don't hesitate to let me know whenever I can return the favor.

(c)   As you know, we've been looking for a new lab assistant ever since the beginning of the year. Your offer to help us find someone was appreciated in itself, but then when you followed it up with the resume which reached me this morning, well, frankly I'm impressed. Your "can do" attitude is remarkable. The answer, of course, is yes—I'll be glad to interview Ms Arlington at any time convenient for her on Monday or Tuesday of next week. If she can cope with the complicated terminology we use around here, we may have an ideal spot for her. Judging from her bio and the grades she received at Columbia, I doubt she'll have any difficulty with the jargon or with some of the newer compounds we're working on. Al, I owe you one. Regardless of how the interview turns out, I'd like you to be my guest at lunch next week. Let me know which day is more convenient for you.

A WORD ON STYLE

Refer to Prior Events

If you are replying to a letter, it is always a good idea to say so immediately: "Thank you for your letter of January 14." An opening like that informs the reader of your reason for writing and also creates a continuing record.

Usually it pays to add something about the subject of the previous letter or phone call: "Thank you for your phone call yesterday in which you inquired about the reasons for the delay in shipping the X-400 ICs."

You may want to refresh your reader's memory about a sequence of events: "We have corresponded (or talked) in the past about your frequent absences from work." That opening sets the stage for the specific content of the letter—which has to do with another absence.

If you refer to prior events in this way, your letters, when looked at in the future, will be self-explanatory. A person reviewing your files will be able to make sense of nearly every letter, without requiring a briefing about the events leading up to each one.

Chapter 4

TESTED LETTERS AND MEMOS THAT RESOLVE PROBLEMS

Among the most challenging letters and memos are those that attempt to correct errors. Unless they are carefully written, they can compound the original error or lead to new problems. The result can be a monster snarl that wastes time and frays the tempers of all concerned. A few basic rules can help you avoid the pitfalls: Keep your temper under control; nothing is gained by making the "opposition" angry. Be clear in your own mind as to the factors involved and what you want to say. Do not dwell excessively on details. Instead, emphasize the desired solution.

When dealing with an outside firm, write to a person, not a box number or a department. If you know someone in the organization by name, direct your letter to that person with the request that, if necessary, it be forwarded to the appropriate individual. If you do not know anyone by name, address the letter to someone with a specific title, for example, Chief Engineer, Design Engineer, President, Research Director, or use a title you think would relate most closely to the subject you're writing about.

Design the format and arrangement of your major points to assist in clarifying the problem. This might involve the use of tabulation and/or underlining to highlight key factors. Emphasize the specifics, such as significant dates, invoice numbers, job assignments, model numbers, catalog numbers, and so on. Make it virtually impossible for the reader to miss the point of what you're saying. Remember, your letter may be passed on to an assistant with only a vague order to "straighten this out." Keep your memo or letter as brief and simple as possible, but not so brief that it takes on a curt or rude tone. Enclose copies of any documents that help to clarify further the problem, but keep the originals. If the nature of the situation is such that you must send originals, be certain you make photocopies and retain them in your files.

Select a basic model from one of the following letters or memos, one that most closely fits the problem you're concerned with, substituting the facts pertaining to your own situation. Bear in mind the "three C's" of effective letter writing. You'll stand a better chance of getting the desired response by making your letter or memo *clear, concise,* and *courteous.*

● *RESOLVING PROBLEMS IN STAFF COMMUNICATIONS*

General Guidelines

Memo 4-1 deals with interdepartmental communications. The writer follows a reliable path in presenting his points. First, he expresses an understanding of how the situation might have confused the person he's writing to. Next, he identifies the cause of the problem. This is followed by his assurance that he agrees with the decision made and includes a brief note of regret about the confusion. Finally, he reinforces his concurrence with the decision and closes by saying he's pleased that everyone is now in "synch," enabling him to end the memo on a positive, upbeat note.

Memo 4-1

April 28, 19—

TO: Barney Jones

FR: Jack Kern

I can understand how my memo of April 7 must have confused you. It was dictated and sent to the Word Processing Department before I learned from Bob Thompson that he had agreed the contract should be changed to include service to the hydro cooler. It now appears we are all in agreement that this change should be made.

Unfortunately, my original memo was sent prior to this news reaching me. I'm sorry for the confusion, but glad that we now agree on all points.

Jack

Alternate Approaches

(a)   I have sent two memos to the Worcester plant about this (see copies attached) and have yet to receive the information requested. You will note that we originally asked for the work and accident record of John L. Baisley.

In response, the Human Resources Department sent me a personnel file on John L. Beasley. As soon as we realized the error, the file was returned to Ellen Wright, and, once again, we requested the file on John L. Baisley. As you know, we're supposed to make these requests in writing; otherwise, I would have been on the phone long before this. Since you work on premises at Worcester, would you please hand deliver this latest memo to Ellen Wright, actually secure the Baisley file from her, and then put it in the mail to me as soon as possible? I'd really appreciate it.

(b)   Recent changes in the organizational structure of our staff required corresponding adjustments in our reporting procedures to management. Now, we've been requested to forward weekly production status reports to the chief engineer's office instead of to the general manager's office. It's interesting to note that this change resulted from a suggestion by someone on our staff, and I'm glad it was approved. We should, however, add one thing to the data on this weekly report. Specifically, the chief engineer should also receive a copy of data compiled by the Quality Control section. I have asked Tom Keating to update this information weekly and attach it to the Production report before it is sent to the chief engineer.

(c)   One of the main topics discussed at recent staff meetings has been the limited feedback we receive from our marketing and customer relations people. This has a direct and negative effect on, among other things, our Product Design people and the new products they are working on. It also affects other facets of our engineering program, and a meeting was held yesterday at which we concentrated solely on this problem. I'm glad to tell you it was a productive meeting. The results may not become visible immediately, but a mechanism is being put into place that will assure regular and useful data provided by Richard Manning (Marketing) and Bill Lawton (Customer Relations). I have attached a list of the questions we'd like these people to answer for us, putting their responses in the form of a report that will be issued monthly. Please review this list carefully. If you feel we've overlooked something, please add it. Return your comments and suggestions directly to me.

- ## *ACKNOWLEDGING PRODUCTION AND/OR SHIPPING ERRORS*

General Guidelines

Letter 4-2 points out that quickly acknowledging an error saves times, avoids protracted correspondence, and assures the reader you are not trying to duck the issue. If you feel the complaint is justified, tell the writer so and express regret for the error. If you determine that the complaint is not justified, you can still convey your concern even if no apology is due. In most cases you'll want to thank the writer for bringing the matter to your attention and, if appropriate, explain how the incident occurred. Make it clear that any error was unintentional and assure the reader that such incidents do not happen very often. If the incident requires follow-up action, specify exactly what you are going to do to rectify the situation and how quickly you expect this action will be taken.

Letter 4-2

APEX MANUFACTURING, INC.
412 Kenyon Drive
Pauley, Kansas 00000

January 28, 19—

Mr. William Mapes
Largo Corporation
220 Ames Street
Coastal, Georgia 00000

Dear Mr. Mapes:

You are correct. Our Production Department did make an error in the outside diameter of the fasteners sent to you on January 12. As a matter of fact, I realize now that two errors were made. You were also entitled to the 7 percent discount, and this was unintentionally overlooked. Please return the entire shipment of fasteners directly to my attention. We will reimburse you for shipping costs. The correct fasteners will be sent to you no later than tomorrow—and I have also instructed our Accounting Department to send you a corrected invoice that includes the 7 percent discount.

I regret the difficulty this caused you, Mr. Mapes, and very much appreciate your calling it to my attention.

Sincerely,

James Watson
Technical Service Manager

Alternate Approaches

(a) If you were able to evaluate the controls we've established to avoid the type of error you brought to our attention, you might agree that this was, indeed, an unlikely occurrence. That doesn't make me feel any better about the problem you had with our X100 model. It does, however, serve as a reminder that human error is something we must try even harder to avoid. You can be sure we want to correct the situation—and quickly. A new rheostat is being sent to you under separate cover. It was checked and double-checked before putting it in the mail. First-class mail, that is. You should have it in the near future. Please return the defective rheostat to me because I want our technical staff to examine it carefully. We will determine just what went wrong and make certain all appropriate people here are informed of the results of this investigation.

(b) It was unfortunate that we were unable to complete the repairs on time, and I apologize for the inconvenience it caused you. We place extraordinary emphasis on a reliable, comprehensive maintenance program. This usually results in our being able to give clients the reliable service they expect from Coastal Jet Maintenance. Once in a while though, particularly with operational equipment as intricate as servo mechanisms, mechanical problems develop that need to be solved before the unit ever leaves our plant. This was the case in this instance, and with the emphasis we place on safety procedures, there was no choice but to delay shipment until we were satisfied the problem had been corrected. Please be assured we will continue to do our best to provide on-time, and reliable, service to you and all our customers.

(c) This is in response to your thoughtful note of May 12. Under the circumstances I am especially grateful for your courtesy. The defective units you received resulted from design changes that were not communicated to all appropriate people in our Production Department. This accounted for several inoperable units being sent to a few customers, and unfortunately you were one of them. Rest assured everyone here is now fully informed, and I'd like you to return the defective units to me at our expense. Please return them directly to my attention. I have also issued instructions to rush new models to you. In addition to correcting the difficulty you described, the replacement units also have several new features which are described in the enclosed brochure. In your case, there will be no cost differential. Thank you again for your letter. Mr. Thurston.

● *CHANGING INADEQUATE COST REDUCTION APPROACHES*

General Guidelines

There are many occasions when a friendly, somewhat informal, request similar to Memo 4-3 can accomplish greater results than is usually the case with a terse, dictatorial approach. Some individuals tend toward memos that resemble edicts, overlooking the fact that a personal touch and friendly tone can often succeed where other methods fail. In some instances, while not necessarily so in this one, the memo might be sent to the employee's home instead of routed through office mail. When implied or direct criticism is involved, it can also be helpful to preface this with reference to something the employee has done that is satisfactory. This can soften the criticism and convey your desire to be objective. There are, of course, occasions when you will want to keep the focus strictly on the action that needs to be taken. In any event, your remarks should be specific and reinforced by facts, not based on general supposition.

Memo 4-3

October 24, 19—

TO: John Harbor

FR: Bill Maxwell

I have been studying the September and year-to-date cost statements for the Apex plant. Overall, your cost cutting has improved since my review of the first six months, but one thing continues to disturb me: the reductions have been in the small cost categories. Please take another look at our budget projections. If you are going to meet the budgeted goal by year-end, costs of finishing supplies and maintenance equipment still need substantial cuts. We discussed this in July. The time for action is fast disappearing. Please send me your plan by October 30 for reducing these costs, in line with budget requirements, for the last two months of this year.

```
Continue the good work on the other items; we can't lose any
ground already gained.

Bill
```

Alternate Approaches

(a)   It was reassuring to note the significant drop in maintenance costs last month, and I want to commend you for this. I also took particular notice of the remarks in your report dealing with recent purchases of additional lab equipment. As you noted, some of this equipment will be needed for new products now in the development stage. It's quite possible two of these products will be delayed because of design problems we've run into, and I've attached a note from the design engineer that goes into more detail about this. Consequently, I'd like you to defer any additional purchases until we have a firm date on when the production cycle will actually begin. We can discuss this in more detail at the staff meeting next Tuesday.

(b)   Your suggestion regarding replacement of control switches for the Model 20 seemed like a good one, until I checked with Bill Hoban in the Production Department. Bill agreed that it offers potential and would probably have put it into effect if not for one crucially important factor. Specifically, his budgetary limits for the current fiscal year. This was a budget that we indirectly contributed to when Bill asked us for our estimates back in November of last year. Now, I'm afraid, it would be somewhat like changing horses in midstream if we changed our minds and decided to use the more expensive switches. Nevertheless, it was a constructive thought, and one that most likely will be put into effect in January. In the meantime, I'd suggest you meet with Bill on this subject and following your meeting send me a note summarizing the results.

(c)   During the staff meeting yesterday, you raised some good questions regarding the cost reduction program now in effect. The incisive and constructive nature of those questions prompted a thought I'd like you to implement within the next two weeks. In essence, I don't think we're sufficiently capitalizing on the ingenuity of our technicians. They are well aware (or certainly should be) of the need to reduce costs, but we haven't really invited or encouraged them to submit specific suggestions for consideration. I'd like you to develop an effective way of ensuring their active participation in this program. Would a luncheon meeting be helpful? Should we invite them to the next staff meeting? Should we send a note on this to their homes? What do you recommend? Please let me have your suggestions within the next day or two.

● *EXPLAINING CAUSE OF ERRORS IN STAFF PROCEDURES*

General Guidelines

As in the case of Letter 4-4, communications regarding staff errors should be preceded by a determination as to who, specifically, was at fault. It is at least possible that the one complaining may have, if only indirectly, contributed to the error. When this is not the case, convey your understanding of the error and explain simply and succinctly how it happened. If appropriate, follow this with a short review of the specific procedural changes you are making to avoid any repetition. The appropriate staff member(s) should also be made aware of your response. Provided a staff procedures manual is maintained, the new or adjusted procedures should be issued in writing to the individual responsible for maintenance of this manual.

Letter 4-4

CARWELL MANUFACTURING CORPORATION
420 Diggs Avenue
Phoebus, Maryland 00000

April 25, 19—

Mr. Robert Smith
Hillsdale Manufacturing Associates
110 Ames Street
Leonard, North Carolina 00000

Dear Mr. Smith:

We do regret that you did not receive the design changes and related equipment you wanted for the Northcon exhibit. As you can see by the copy of your order which is enclosed, someone inadvertently wrote down the wrong stock number.

You will be happy to know that everything you need was shipped by air freight within an hour after your telephone call to me. You should have it tomorrow. The other material, if you cannot use it, should be returned to us for credit to your account.

Cordially,

Roy Brown
Service Manager

Alternate Approaches

(a)   Last month a new policy was adopted for handling requests from outside individuals or companies for permission to reproduce or otherwise use material from our publications. This policy was circumvented last week, and I want to make certain it doesn't happen again. There were mitigating circumstances because the individual responsible had recently been assigned to our division and had not received his staff procedures manual. From now on, it is imperative that staff supervisors provide new personnel with this manual the first day they report to work and, whenever possible, prior to their first day on the job. A slip should accompany this manual. You can draft the message the slip contains (send me a copy), but in essence it should require the new employee to indicate he or she has read the manual, agrees to follow the procedures indicated, and lists any questions he or she might have.

(b)   The reasons for our safety program are obvious to all of you. The benefits to be had by following these provisions are also obvious. Yet we have had two accidents on the loading platform in the past few days, injuring three employees and damaging our overall staff performance. This, I hope, was a momentary lapse in a safety program that was proving to be quite effective until this happened. I'd like the supervisors of the employees involved to meet in my office on Monday of next week. Please bring with you a written description of the causes that led to these accidents, and your recommendations on how we can avoid any repetition in the future.

(c)   The quality control supervisor, quite correctly, brought to my attention several products recently that he thought were defective. His reasoning was that the tabs on the edges of the RX chassis were missing and had somehow been omitted during the production process. You know, and I know, that design changes led to the elimination of these tabs, but no one arranged to pass the word along to the Quality Control people. Unfortunately, Bill Manning (the supervisor) stopped the assembly line until this matter could be cleared up. Thus, we lost a significant amount of time and money because of the breakdown in communications. Please (and make certain this is passed along to your people), do not institute design changes without following our established policy of notifying, among others, the quality control supervisor.

● *ESTABLISHING CORRECT PROCEDURES FOR REIMBURSEMENT*

General Guidelines

Letter 4-5 reflects a situation that occurs most often in larger organizations, particularly when there are branches reporting to one headquarter's office. The matter of reimbursement and payment for services rendered lies close to the bottom line in any company, so letters and memos on this subject take on added importance. This is another area in which procedural details are sometimes not known by one of the parties involved. Letters of this type need to explain why payment has not been processed more promptly, and outline clearly the action required before reimbursement is arranged.

Letter 4-5

RIVERTON CORPORATION
460 Walton Lane
Cartwell, California 00000

December 4, 19—

Ms Ellen Arthur
480 Bentley Road
Dawson, Michigan 00000

Dear Ms Arthur:

I am sorry we cannot write you a check from our local plant for your past-due bill No. 278-089789, as you requested.

We received the copy of this bill for lab equipment on Wednesday, November 21. It has been matched with our purchase order and company procedure requires that we send it to our headquarters plant in Detroit for payment.

Although payment is overdue (because we did not receive the original copy until yesterday), you may be sure our headquarters office will issue payment shortly after our request reaches them. This procedure normally speeds payment in practically all instances. Unfortunately, this is one of the exceptions.

You should receive payment within the next week or ten days.

Sincerely,

Robert Benton

Alternate Approaches

(a)   We received your letter of May 12 and understand your concern regarding payment for consulting services performed from April 4 through April 11. The matter would normally have been taken care of within ten days following completion of your assignment, but we have not received your response to our letter of May 1 (copy enclosed). I believe you may have been informed of this procedure earlier, but I apologize if you were not. In essence, we require that all bills of this nature be accompanied by an itemized listing of work performed, showing the amount of time spent by the consultant on each item listed. When this has been received and approved by Lawrence Green, chief engineer, we will arrange for payment just as quickly as possible.

(b)   In reviewing our client records we find that your account occasionally exceeds our terms of sale and now has a past-due balance of $1,880.00. It may be that the standard 30-day credit arrangement is not the best one in your particular case. If our records are in error, we would appreciate hearing from you so that any correction necessary can be made. Also, if you feel that a different type of payment procedure would be more suitable for your organization, we would be willing to discuss this. In any event, would you please let us have your response in the near future. We are pleased to have you as a client and look forward to continuing our pleasant and mutually beneficial association.

(c)   The Apex Company located at 122 Kenyon Avenue in Omaha has placed an order with us for an extensive variety of hydraulic lift equipment. As a means of establishing credit, they provided us with a list of several firms which they have done business with. Since your organization was on this list we would appreciate any comments you can give us concerning this company's practice in meeting financial obligations. Please be assured your response will be held in strict confidence. This letter is also in response to the request made by William Atkins of your organization who, during our recent telephone conversation, requested that we put this inquiry in the form of a letter to you. Thank you very much for your cooperation, and in the interests of all concerned, we'd be grateful if you could send your comments within the next few days.

● *IMPROVING METHODS OF PRODUCT SHIPMENT*

General Guidelines

When correcting an earlier action or issuing related instructions, Letter 4-6 illustrates two rules to bear in mind. The first, making your position in the matter clear, is obvious. The second is embodied in the style and approach of your letter. Here you want to be careful to avoid an imperious tone or manner that might be viewed as condescending. State the facts simply and briefly. If an explanation of the original and improper action is required, provide it in a factual, concise way. Then quickly move on to the corrective action you are taking, conveying this with a positive and courteous tone. Little is to be gained by an endless recital of the factors leading to both the problem and your solution, so brevity could be termed a third rule to bear in mind with letters of this type.

Letter 4-6

BRINKLEY INTERNATIONAL MANUFACTURING CORPORATION
8446 Main Boulevard
Carson, Washington 00000

March 28, 19—

Mr. William Bruno
Chief Engineer
Forster Machine Tools, Inc.
80 Market Street
Ames, Ohio 00000

Dear Mr. Bruno:

We shipped your order No. 84222 for three air valves air express rather than parcel post because we assumed time was a critical factor.

Because our assumption was in error, a credit memo is being sent to you today for the difference in delivery rates. We regret the misunderstanding, but assure you it resulted from our desire to be of service to you.

Sincerely,

Alternate Approaches

(a)   You are justified in being annoyed by our blunder in returning the unordered equipment you sent to our warehouse. Please understand that this error was completely unintentional. It would really help a great deal, however, if when you return equipment, a note is enclosed to me or the engineer you've been working with, stating the reason for the return. This will enable us to make certain proper credit is applied. In this instance, you will receive immediate credit for the returned equipment, and this will include credit for the shipping charges involved. Again, we regret the inconvenience to you. We do value your business and your friendship.

(b)   Many thanks for your letter of July 17 describing damage to the PC boards you ordered. Bringing this to our attention quickly has enabled us to correct the problem quickly. A replacement order was shipped to you today. We do regret the damaged shipment and the inconvenience it caused you. The reason can be stated briefly. We installed a new conveyor system in our warehouse, and although it is now working beautifully, we did have an occasional setback when it was first put into operation. Please contact me directly if you do not have the new shipment within the next few days.

(c)   Our discussion of current shipping procedures was helpful to me. I realize that as shipping supervisor you face a variety of problems each day and some of them take time to solve. With respect to one of the things we discussed, there is a request I'd like to make. I hope you agree it is worth careful consideration. General Laboratories, Inc., is one of our most important clients. We've temporarily assigned two of our staff members to General to assemble and calibrate the lab equipment it recently purchased. I suspect General may either return or request an exchange for some of this equipment. If this happens, I would appreciate it if you would let me know immediately. It could have a bearing on other engineering services we would like to perform for this account.

● *APPLYING PRESSURE TO EXPEDITE*
   COLLECTION PROCEDURES

General Guidelines

As in the case of Letter 4-7, the obvious purpose of a collection letter is to collect payment for goods or services rendered. In an ideal sense, the writer should attempt to accomplish this while continuing to retain the goodwill of the debtor. The most important thing, of course, is to identify exactly what is delinquent. One way to increase the likelihood of getting a response is to include a prepaid, self-addressed envelope. If you do, your letter should make mention of this. We are assuming here that you have gone through the customary initial steps of collecting payment. If there has been no response, or promises have been made but not kept, it becomes necessary to use a stronger, more insistent tone. An adaptation of Letter 4-7 may be sent (consider using registered or certified mail) just prior to turning the matter over to a collection agency, or your attorney. Also, note Letter 3-4, which deals with comparable situations before they reach the "collection" stage.

Letter 4-7

ARCWELL CORPORATION
458 Bernita Terrace
Bangor, Idaho 00000

June 14, 19—

Mr. Edward Nelson
52 House Street
Hampton, Illinois 00000

Dear Mr. Nelson:

On May 12, May 24, and June 6, we wrote to you with copies of the
bill for lab equipment that was sent to you on April 5. We have
received no response.

Until recently, we assumed that failure to resolve the matter was
an oversight and that our letters would bring an appropriate response, enabling us to continue to make laboratory equipment
available to you. I believe you are aware of what an "appropriate"
response would be, but it appears that you have not even extended
the courtesy of any response at all. The need for full payment, or
return of this equipment, has become urgent. Would you please
make certain this matter is resolved quickly.

If every item, as indicated on the enclosed list, is not returned
by June 20, or payment in full is not received by that date, our
attorney will take whatever action is necessary to ensure a satisfactory resolution of this matter.

Sincerely,

Alternate Approaches

(a)   You've probably had experiences with firms that do not pay their
bills as promptly as you would like. Some of them need just a gentle reminder, others may require several reminders, and a few do not respond
until they receive an ultimatum. We genuinely regret that our situation with
your organization has reached a critical stage. The enclosed chronology

of events lists the supplies you have purchased since May 1, and the attempts we have made to secure payment. The facts supply evidence that we have not only been patient, but have made every effort to develop a payment schedule that would be realistic and reasonable for both of us. Our attorney has been told to begin litigation proceedings if payment in full is not received by July 1.

(b)   There is a procedure we have access to that would enable us to bring about a resolution of your outstanding debt to our company. In your best interests as well as ours, we'd like to avoid this. It relates to our contract with the local credit bureau, requiring us to report all past-due accounts to its Collection Department. Of course, taking this action would jeopardize your credit standing in the community, and this could exert a far-reaching and harmful effect on your business activity in this area. The Better Business Bureau would also be duty bound to report this to anyone inquiring about your firm. You can help us avoid the necessity of taking this action, and in your own best interests, I would urge you to do so. Unless we receive payment in full by the end of this month, we will have no recourse but to take this approach.

(c)   Our Legal Department has instructed us to send you the enclosed Notice of Delinquency. We had hoped to avoid this, but the fact you have failed to respond to our earlier letters leaves us no recourse. Even though we are at the point of court action on this matter, I still find it difficult to believe you would want this to happen. There is a way to settle the matter once and for all and avoid the danger of having a poor credit rating affect your business. Obviously, the way to do this is by issuing a check to us for payment in full. I sincerely hope you will do this in the best interests of all concerned. If you do, and the check is received no later than May 15, we will take no further action. If you do not, then the next step in this matter will be taken by our corporate attorney.

● *RESOLVING QUESTIONS AND COMPLAINTS FROM CLIENTS*

General Guidelines

Letter 4-8 illustrates one way to begin a letter that, essentially, says "no" to a client's request. It is sometimes helpful to open such a letter by agreeing with the writer on a particular point. Another recommended opening statement might express your appreciation for the client's inquiry. Even if your answer is still no, it will give you "an opportunity to explain" to the writer your position in the matter. Merely to say "it is against company policy" or "for various reasons we cannot comply" is tantamount to ignoring the reader, and inviting more critical correspondence in the future. When you are able to grant the writer's request, or provide satisfactory answers to questions raised, capitalize on this by reinforcing the client's relationship with your organization. Convey the implication you're glad to be of service because you want that relationship to continue.

Letter 4-8

GOODWIN ELECTRICAL PRODUCTS, INC.
489 Herman Avenue
Milwaukee, Wisconsin 00000

March 7, 19—

Ms Eleanor Gerald
412 Bernita Drive
Chicago, Illinois 00000

Dear Ms Gerald:

We agree that an "automatic" timer that does not respond to precise settings is useless. Since your letter involves a technical repair, it was referred to me.

As you know, we guarantee trouble-free operation of our timers for a year from the purchase date. This guarantee applies to manufacture-related defects.

Your timer has been examined by our Technical Service Department, and they report deep scratches and a dent near the setting mechanism. This indicates physical damage to the case, which is not covered by the guarantee. The dent is deep enough to touch the main spring regulator when it expands, for example, in warm weather. This can cause the occasional malfunctions you reported.

Although not covered by the corporate guarantee, we will repair your timer at factory cost and with a warranty that covers this repair for three months. The cost to you is indicated on the enclosed repair schedule.

If you wish to have the repairs made, please return the enclosed
card directly to my attention and attach your check. We will repair your timer and return it promptly.

Sincerely,

Karl Dietrich
Technical Service Manager

Alternate Approaches

(a)   Your letter regarding credit for the gear housing we made for you
reached my office this morning. Although we realize that machinery rebuilding plans are sometimes changed at levels above that of the purchasing
agent, we regret that we cannot accept return of the gear housing for credit.
The reason, quite simply, is that it was cast to your specifications and there
is just no way we could market it to one of our other customers. This,
unfortunately, is the present case, but if by some remote chance we get
another inquiry for this type of casting, you can be sure I'll get in touch with
you immediately. I wish we could do more for you, Mr. Jackson, but under
the circumstances, I hope you'll understand our position in the matter.

(b)   I'm delighted to tell you that your paper dealing with polymer compounds is one we'd like to publish. It is unique in several respects, and we
believe it will offer special appeal to the large audience we reach via direct
mail. As you may know, it is our practice to secure reviews from three
experts in the subject area covered by a particular presentation. We are in
the process of doing that right now. It usually takes only two or three weeks
to get these reviews, and as soon as they have been received, I'll be in touch
with you again. In the meantime, please review the enclosed terms of publication, and if you have any questions, let me know. While we have no way of
knowing what the reviewers' comments will be, I hope their overall reaction will be as favorable as ours. If so, a formal contract will be sent to you,
and this will be similar to the "Terms of Publication" I have enclosed.

(c)   Please refer to your Claim No. 44214. Our records show your bill of
lading #2882 was signed over four months ago, on January 12. The date of
your claim is May 30. As you will note, Section 2(b) of the Bill of Lading
Contract Terms and Conditions stipulates that claims must be filed in
writing within three months from the date of delivery. In light of these
requirements, we are sorry we cannot honor your claim. We wanted to
make certain about the dates indicated and, as a result, checked all possible
sources for a prior filing. Despite this careful search, we could find no

record of an earlier claim related to this shipment. You may be certain, however, that if you can provide tangible evidence of an earlier filing, we'll be glad to reconsider this claim. Thank you for your letter, Mr. Cartwright. I sincerely wish I could be of greater help to you.

● *ENSURING CORRECT USE AND APPLICATION OF PRODUCT*

General Guidelines

A point made indirectly by Letter 4-9 is that it does little good to consummate a sale to a particular client and then have the client ignore installation (or application) instructions. The situation is not uncommon, and the way in which you handle the matter will determine whether the individual eventually becomes a satisfied client or not. It is all too true that many customers follow the old saw "As a last resort, read the instructions." When the client chooses or neglects to follow clear guidelines provided with the product, the manufacturer must use a deft touch in dealing with the results. Responding with a cold and obvious form letter does more harm than good. When adapting the sample letter or one of the "Alternate Approaches" provided, use a courteous, personalized approach. Convey your genuine concern and desire to help, even though the fault does not lie with the product itself. Of course, if it becomes necessary to deal frequently with letters of this type, the guidelines themselves become suspect.

Letter 4-9

RYAN AND CONNER, INC.
1220 Davis Street
St. John, Minnesota 00000

March 7, 19—

Mr. Thomas Wilson
1540 Maxwell Street
Washington, D.C. 00000

Dear Mr. Wilson:

Your recent inquiry dealt with the technical aspects of applying our Thickwall product; consequently your letter to our corporate president has been referred to me.

We're sorry to learn that your application of Thickwall was not completely to your satisfaction.

Applying Thickwall is a highly specialized job. That is why we request all our dealers to make this part of the warranty very clear to each customer. Because of the specialized techniques involved, it must be applied by someone who has the technical expertise necessary to complete the application satisfactorily. We understand that Bob Johnson, the salesperson at your local dealer (Appley Building Supplies), recommended application by John Sanders, a specialist who works with Appley Building Supplies. Once again, the warranty states quite clearly that Thickwall must be applied by a specialist.

I have talked to Bob Johnson, and he suggests that John Sanders inspect your house and make recommendations for correcting the problem. If you and Mr. Sanders can agree on this, then perhaps he can make the necessary repairs at reasonable cost to you—or perhaps suggest how you can make the repairs yourself.

Please talk to Bob Johnson at Appley Building Supplies. He assures me that he's anxious to work out a satisfactory arrangement for getting the application of Thickwall corrected.

Sincerely,

Mitchell Connors
Technical Service Manager

Alternate Approaches

(a)   It's unfortunate that you had a problem with the installation of our safety barrier strips. As you say, they are designed to prevent short circuits, not cause them. We were as puzzled as you were regarding the source of the problem—until they were returned to us. They arrived shortly after receipt of your letter, and the cause immediately became obvious. I have enclosed a sample of one barrier strip that represents the type we sent to you. Please note carefully the other barrier strip, which is one of those you returned. The insulation strip should extend slightly beyond the base, but a tiny portion had been removed from all the strips you returned. This may have been done because of space limitations on the PC board, but the action caused a short to occur between the strip and the metal chassis. We'll be glad to send smaller barrier strips, just as soon as you send us the correct dimensions.

(b) Thank you for writing to us about your application of our mercury switches. It is our policy to include detailed specifications of the switches themselves, along with application data similar to the material I have enclosed. I assume (and I believe it is a correct assumption) that your staff received this material with the mercury switches themselves. Please note the electrical parameters: "Rated 5 amps at 125 V AC." When your representative explained your use of the switches, we realized the voltage involved probably exceeded the limitations indicated in our printed material. Our Engineering Department would like to help you resolve this. If you will send me a schematic of the circuitry you're working with, we'll try to suggest a more appropriate switching device that will serve your purpose.

(c) I appreciated the detailed information provided in your letter, regarding the Inline coax amplifiers you purchased. Because of this helpful data, we were able to identify the cause of the difficulty. While it is quite true that these units can be placed anywhere in the signal path that is most convenient, they are designed for coax cable. The latter was not the problem, because you indicate coax cable is used in your installation. The primary cause of the difficulty is that these Inline amplifiers are strictly for UHF/VHF/FM transmissions within the range of 5 to 900 MHz, using 75-ohm input/output. Thus, they really could not be expected to perform satisfactorily with the AM application (150–2,194 kHz) you had in mind.

● *CORRECTING IMPROPER FULFILLMENT PROCEDURES*

General Guidelines

Requesting corrective action, as illustrated by Letter 4-10, requires emphasis on specific facts if the reader is to know exactly what you want done. Rather than dwell on circumstances leading up to the request, it is usually best to focus on the present situation and a clear outline of those changes that are required. Difficulties can arise because the writer may not be aware of precisely what action the reader may be able to take. Or, possibly, there are various alternative courses of action to be considered. In any event, state the facts concisely, and wherever possible, include your preference as to the remedial steps that should be taken.

Letter 4-10

NEWPORT TECHNICAL PUBLICATIONS
4488 Michigan Boulevard
Atlantic City, New Jersey 00000

April 25, 19—

Mr. George Ryan
Amtex Printers, Inc.
9022 Vine Street
Elson, South Carolina 00000

Dear Mr. Ryan:

The paper stock you sent us would not be suitable for the job we
have in mind, even though you could furnish it at the same price
as the cheaper one we requested. Its extra weight would increase
postage to such an extent that the total cost of the training
manual would be considerably higher.

I am afraid we will have to count your company out of this particular job because production time allowed us is so short. In
the meantime, a bid has been submitted by another printer, and he
has our stock available. He also has the special typefaces that
would be necessary for some of the chemical formulas.

Thank you for your efforts, in any case. I'll let you know when
another job is in the offing, because we do enjoy dealing with
you.

Cordially,

Jack Roes
Engineering Staff

Alternate Approaches

(a) I pointed out several weeks ago that the finish polishing section has
become a bottleneck in our assembly line. From my vantage point it's
difficult to pinpoint the reasons for this, but it is not difficult to observe the

results, or estimate what it's costing us. We simply are not filling orders as promptly as we should, and the time has come for firm, corrective action. I will not presume, at this point, to tell you what action needs to be taken. You not only have the responsibility, but you are in a much better position than I to determine this. Would you please give this top priority and let me have your recommendations no later than Monday of next week.

(b)   R. L. Manufacturing is one of our more profitable accounts. Until recently we served them 100 percent. During our retooling program, and because of delays in fulfillment, R. L. gave a trial order to our competitor, Smith Manufacturing. Now, Smith has about 30 percent of R. L.'s business. Two problems contribute to this: R. L. has repeatedly asked us not to underrun orders because they have specific locations that must be served. We continue, however, to ship from both under and overruns. Second, they pick up most of their orders from us. Yet, all too often they make an appointment to pick up at a specific time, and then have to wait as long as four hours to be loaded. Idle driver time costs R. L. as much as on-the-road time. What are you doing to correct these problems? How can I help? I would like your reply by July 20.

(c)   After reviewing your production forecast for the next three months, I can see trouble ahead. You may feel there's some relationship here to the budget matter we discussed. However, when you had the opportunity to make changes in the budget, you elected to stay with the one put together by your predecessor. The largest part of this problem will occur in March, when we'll need to ship the new presses to Carter Manufacturing. Your staff forecasts a drop in production during February of 190M units. At that daily rate, and considering March has three additional working days, the real crunch will occur at a time when we can least afford it. I must look to you for a solution, Tom. This is your primary area of responsibility, and relates so closely to our order fulfillment capability that we can't afford to delay action on this. What do you recommend?

● *SECURING APPROPRIATE CREDIT TO OFFSET ERRORS*

General Guidelines

As contrasted with some of the earlier entries, Letter 4-11 deals with the situation after an error has occurred. (See, too, Letter 4-5.) Mistakes are inevitable in any enterprise, and at least two forms of action should follow an error of this type: correction of the procedure or mechanism that caused the problem in the first place and appropriate action to compensate or reimburse the client for any loss he or she may have suffered. As with other letters and memos dealing with a comparable situation, the tone is just as important as the content. It may be necessary, in some cases, to convince the offending party that an error has indeed been made. This should be done with facts, presented clearly and objectively. The action you want taken to correct the situation should be presented with equal clarity.

Letter 4-11

CAMDEN IRON WORKS, INC.
1214 Ascot Avenue
Huntington, New York 00000

August 5, 19—

Mr. Louis Cote
Rockwell Milling Corporation
400 Kent Avenue
Bentley, Oregon 00000

Dear Mr. Cote:

This letter provides a synopsis of the facts relayed to you during our telephone conversation of yesterday. On July 18 your Cincinnati mill shipped a total of 100 3-inch finishing rollers to our Huntington plant, relative to our order JU-0000. This order, as you are now aware, requested only 10 rollers.

Fortunately, we were able to resolve the disposition problem partially by having our Huntington plant accept 20, the Lexington plant accepted 20, and Charleston took 20, with the remaining 40 being shipped back to your Knoxville plant.

I have enclosed copies of your tally sheets, our Bill of Lading for the shipment to your Knoxville plant, along with the freight bills for the shipments to our Lexington and Charleston plants to substantiate our claim. As outlined on the attached recap sheet, our claim includes the freight charges to Lexington, Charleston, and Knoxville.

Please review this as soon as possible and issue the necessary credit to our Huntington plant. I believe you are now in posses-

sion of all the pertinent facts related to this situation. If you
require additional data, please let me know immediately.

Sincerely,

Thomas Kiley
Plant Manager

Alternate Approaches

(a)   Data enclosed with your recent shipment of variable attenuators
indicated customers must notify you before returning any of your products.
This letter provides that notification. We refer to Order #44882, copy enclosed. You charged us $422.00 for this shipment, and, as you will note, our
original request (copy also enclosed) was for 50 300-ohm variable attenuators. Your firm incorrectly filled this order with 75-ohm attenuators. We
await your instructions on the return of this shipment and a full refund of
the payment we made (our check #6821). Since time was a crucial factor, we
have since purchased the 300-ohm attenuators from another source.

(b)   The steel tubing we ordered on March 15 arrived on March 21, and
I want to thank you for fulfilling our request for a rush shipment. Perhaps it
was the urgency of the situation, and your desire to help us out on this, that
caused the problem. The problem being that 100 of the 5′ lengths were
provided in 16-gauge steel, instead of the 20-gauge steel we ordered, and
were not swaged at one end. The remaining supply conformed to our
original specifications and present no problem. However, we have no need
for the 100 lengths we refer to and have no recourse but to return them to
you. If you can replace them using the 20-gauge steel we ordered, fine.
Please do so as quickly as possible. If you cannot do this by April 4, we will
have to go elsewhere. In that event, please let me know when the appropriate credit will be issued to our account.

(c)   Thank you for your courtesy during our telephone conversation. This
letter will summarize our discussion. The RF modulators we ordered on
May 6 were to be used to convert line-level (baseband) signals from satellite
TV receivers to VHF for reception on standard TV sets. When we spoke to
Mr. Allen on your technical staff, he frankly expressed some uncertainty as
to whether your RF modulators would work with our equipment (which we
described to him in detail). However, he felt the only way to make certain
was to try them. We did, and unfortunately, with unsatisfactory results. I
would appreciate it if you would issue the necessary credit. You were kind

enough to send the units via Federal Express, and if you wish us to do so, we'll be glad to return them the same way. Thank you for your cooperation.

●　*ISSUING INSTRUCTIONS THAT HELP AVOID FUTURE ERRORS*

General Guidelines

Letter 4-12 represents the type of letter that no one likes to write. Yet a situation involving product returns (or dissatisfaction with services) will sometimes represent an opportunity in disguise. At the very least, it offers a way to lessen a problem caused by a defective product, and perhaps eventually enable you to make a friend out of a dissatisfied client. Letters of this type also reflect your desire to help the client overcome an unpleasant experience with others in your organization. The tone and content of your letter should be designed to do this. If successful, the aftermath can be beneficial to all concerned. Letters of instruction, whether to staff members or to clients, must frequently deal with errors that have been made or provide a means to avoid errors that could occur in the future. Either way, they represent opportunities to clarify, and correct, circumstances that could be detrimental to the parties involved.

Letter 4-12

WORCESTER INDUSTRIAL PRODUCTS, INC.
448 Oakley Boulevard
Bentley, Illinois 00000

April 24, 19—

Ms Beverly Morton
Design Engineer
Alderney Manufacturing Co., Inc.
26 Keystone Drive
Oneida, Virginia 00000

Dear Ms Morton:

I don't blame you for being annoyed about our double blunder with the formula for reduced quantities of Tri-X. Please allow me to

apologize. I have no excuse for the errors, but if you should
ever again have to return something we sent you (and I sincerely
hope you will not), please include a letter or note addressed to
me or my assistant, Bill Watson. This will assure careful attention and prompt handling.

Regarding the current mess, please have the shipment returned to
us *marked for my attention*. I'll see that you are promptly cred-
ited and this will also include shipping charges.

Again, let me say how sorry I am for the inconvenience caused you.
We do value your patronage and our long-standing relationship.

Sincerely,

Alternate Approaches

(a)   The purpose of this letter is to follow up on our phone conversation
of July 11 and your letter of July 17. The primary difficulty appears to have
started earlier this year when some operations were discontinued at the
Worcester plant. At that time we were requested to route some reports to
San Jose that previously had gone to Worcester. We have now issued instructions to correct the problem pointed out in your letter. I believe this
will eliminate the mailing problem. I should mention, however, that we
have found it extremely difficult to maintain an up-to-date listing of who
should get which report. Would you please send me a copy of your current
list, and instruct your assistant to update this on a monthly basis, sending
each updated list to me no later than the first week in each month.

(b)   I would like to ask a favor that could help us avoid any repetition of
the error referred to in your letter. We shipped your order #86642 for six air
valves via air express, and not via United Parcel, because we assumed time
was a critical factor. As a matter of fact, I'm surprised we did not do this on
the preceding order. The reason is one that became apparent as soon as we
reviewed the order in question. Specifically, your Purchasing Department
indicated in large capital letters "RUSH" on this order as well as on the one
preceding it. Normally, we respond to this by sending the shipment by air
express. In the future, if you will instruct your Purchasing Department to
omit the word "RUSH" on orders you want shipped via UPS, we'll be glad
to see that this is done. Thanks very much, Mr. Carter.

(c)   This will acknowledge receipt of your recent letter concerning the
problems you described with the Armstrong power stapler. We regret the
circumstances that prompted you to write. To make certain these problems

receive prompt attention, we have forwarded a copy of your letter to the local distributor in your area and another copy to our Akron district office for their review. I have also enclosed a copy of our instructions to both offices. A representative from the Akron office will contact you in the near future. If you have problems with any of our products in the future, and I sincerely hope you do not, faster action can be secured by writing directly to Mr. Robert Reese, who is our technical service manager in the Akron district office.

● *SECURING CORRECT DATA ON REPAIRS TO PLANT EQUIPMENT*

General Guidelines

Memo 4-13 relates to an ongoing maintenance program for plant equipment and the need for factual information on which to base repair time and cost estimates. Depending on the circumstances, inquiries of this type may need to be directed to staff members or outside contractors. In either event, the information required should be put in sharp focus, with a clear indication as to why and when it is needed. Even if one person is in charge of several locations where the particular equipment referred to is located, it is usually a good idea to send copies of your inquiry to the local supervisors. Reason: they will be aware of the request and can more quickly develop the requested information, sending it to the manager who has primary responsibility.

Memo 4-13

March 21, 19—

TO: Branch Managers

FR: Keith Caldwell

The Johnson hoists at many branches have fallen into disrepair, and many are no longer used for the reasons we discussed earlier.

I would like to evaluate the status of this equipment at each branch to determine the cost of repair, how much help you may need from Central Engineering, and what performance and cost-saving opportunities are available.

By Monday, April 17, I would like replies to the following questions from each of you.

1. Are the hoists operational? If not, why not?

2. What repairs have been made in the past six months, and with what results?

3. What repair work is presently needed? What is your estimate of the time and cost involved?

4. Aside from repairs, what additional improvements would you recommend with the present system?

John Harvey is available at Central Engineering to help you respond to these questions. You can reach him on Ext. 2457.

Alternate Approaches

(a)   In reviewing our major construction project list, we should not overlook the need to correct existing problems with some of our plant equipment. A good case in point has to do with the matter of sound enclosures for the hammer mill. We've had a number of changes here in personnel, and this has resulted in some concern as to whether my files on the noise citation by the state are complete. I believe your office maintains up-to-date files on matters of this type. Would you please have someone make a search on this and send copies of the original citation to me, along with any relevant data that will help us make a decision on changes to be made. I believe there's a time limit on correcting the situation, so I'd like to have this information no later than Monday of next week. If there are any questions, call me on Ext. 2457.

(b)   As of March 1 wc will no longer be able to use the present motor housings for machines in the Toledo plant. The new AC/DC motors will be delivered next week, and we expect installation to be completed within two or three days. This means the housings must be replaced immediately after arrival of the new motors to avoid any costly delay in resumption of production. Please do the following: let me have your estimate on time and cost in making the changeover, your recommendation as to how we might be able to make profitable use of the present housings, and if you feel we have no practical use for them, how we should dispose of them. In the latter event, who do you think we might sell them to, and for how much? I'd appreciate your response no later than February 15.

(c)   Per our discussion this morning, the bins in the lamp assembly plant have insufficient storage capacity and should be replaced as soon as possible. I would like you to go over this carefully with your staff in order to come up with appropriate dimensions for the new bins. This should be done quickly. Bear in mind that this capital investment will be a substantial one, and your estimate should be tied in with the long-range plans we discussed at the manager's meeting last week. What we do now will determine whether we have sufficient parts storage space for the next two years. Also take into consideration the new products now on the drawing board, scheduled to go into production in September. I'd appreciate having your recommendations within the next two weeks.

● *CANCELING ORDERS AND REQUESTS*

General Guidelines

Sometimes a cancelation of the type contained in Letter 4-14 becomes necessary. The circumstances vary, but the end result is usually accompanied by frustration for both parties. A brief, courteous letter explaining your position will usually do the job, but more complicated problems that may lead to litigation should be taken up with your legal advisor beforehand. Obviously, the terms stated in an order or contract should be carefully reviewed before issuing a cancelation. Once you are certain of your ground, the letter should make the cancelation itself very clear, and just as clearly state the specific reason for the action. The tone should be straightforward and as tactful as possible. You may want to do business with the firm or individual at some point in the future.

Letter 4-14

PREMIER ELECTRONICS COMPANY
455 Gainsboro Avenue
Talusa, South Carolina 00000

March 26, 19—

Ms Helen Brown
Ames Technical Products, Inc.
800 Martin Boulevard
Winchester, Vermont 00000

Dear Ms Brown:

Since you are still not able to send us the material listed in
our Purchase Order #24689, we are forced to cancel this order.
As you know, we needed these items to complete a shipment to our
Australian account.

The customer in Sydney informed us that they can no longer wait
for parts from the Ames Corporation. Consequently, the shipment
is being sent without your merchandise.

We are sorry this cancelation is necessary, but we really have no
other choice.

Yours truly,

_____

Alternate Approaches

(a)   Confirming our telephone conversation of yesterday, we find it necessary to cancel our data processing service contract with you as of March 1. The reason stems from a decision to relocate our Hillsdale office to the main headquarters building in Albany. We recognize that the terms of this contract will require payment through May 1, and you may be certain we will honor this provision. Prior to issuance of the check, however, we would appreciate return of the documents (including the technical manuals) we made available to your staff. All this material should be sent directly to my attention. Your cooperation is appreciated, and I hope we will have an opportunity to utilize your service again in the future.

(b)   The order we had for water distillation units has been canceled, so we no longer need the equipment you were going to provide us with. Frankly, we are disappointed by this turn of events too, and it is at least possible we'll be able to locate another client who will want to purchase the units in question. If so, I would be glad to get in touch with you about this. Since shipment of the equipment had not been made, I assume this notice of cancelation will be sufficient. Per a call to your Production Department yesterday, I was relieved to know you had not started work on processing our earlier request. Thank you for your understanding in this matter. I do hope we'll be able to do business together at some future date.

(c)   We have decided not to follow through on our tentative plans for expansion of the Salem office. This comes as a personal disappointment because we do need the additional space, and I regret that budgetary considerations have forced a postponement. We will, of course, see that you are reimbursed for your efforts to date, but I wanted to notify you of this decision as quickly as possible. If you will send me an itemized list of expenses you've incurred during the period March 12 to and including April 6, I'll seek the necessary approval from our district manager so that payment can be made. As indicated earlier, this is indeed a "postponement," although I have no specific date as to when work on the project will be resumed. You may be certain I'll notify you just as soon as a decision is made on this.

● *UNCOVERING PRIMARY CAUSE OF PRODUCT COMPLAINTS*

General Guidelines

Inquiries or requests similar to the one in Memo 4-15 are useful in one particular respect. The results can make a solid contribution toward improvement of a product or service. This type of communication need not be limited to an exchange between manufacturer and customer. Employees can help uncover the reason for customer complaints, although, rather surprisingly, rank and file staff members frequently do not receive feedback from departments that deal directly with these complaints, for example, the Customer Relations Department. There are also instances when customers do not consider a minor defect important enough to justify a written complaint. The "Alternate Approaches" section provides a method for securing helpful information in situations of this type. Few organizations can afford to ignore even "minor" complaints.

Memo 4-15

March 21, 19—

TO: Engineering Staff

FR: William Hiller

In recent months we received a number of product complaints that
could not be traced back to specific production dates and staff
members. One way to resolve this would be to stamp or print the
factory job number on each item. In doing this we would also
separate the product line coming out of our plant from jobs that
are manufactured for us by outside firms.

Please consider this idea and let me have your recommendations no
later than April 4.

William Hiller
Chief Engineer

Alternate Approaches

(a)   This request stems from a number of critical letters we've received
lately on our stainless steel cleats. The complaint in each case refers to a
lack of holding power. We have enclosed samples of the cleats in current
production (#488) and samples of a redesigned model (#448A). Would you
analyze both sets for tensile strength under both steady and alternating
tension conditions. The enclosed specification data will be useful to you
when conducting these tests. Please send us a comparison of the test results
no later than March 15. We are anxious to make a decision on production
changes relative to model #488A. I'd like you to call me as soon as you have
the results; then follow this up with your written report.

(b)   This confirms the requests made during our discussion this morning. We've received 36 complaints in the past month from clients who
ordered plastic containers produced in our Southside plant. In each case,
the complaint had to do with loose lids. It has reached the point where
several customers have specifically indicated their orders should be filled
by the Worcester plant. Worcester's production is geared to the same design

specs, but somehow they have managed to avoid this problem. I would like you to begin work immediately on a report covering the conditions causing this problem at Southside, the steps you're taking to overcome it, and those steps you will take to avoid any recurrence in the future. There should be no need for additional equipment. We have checked this already, and both plants are operating with identical equipment.

(c)   Thank you for purchasing our product X. Even though considerable planning went into the development of this product, responding to the needs of our clients is a never-ending process. The fact we must constantly be alert to the changing requirements of the people we serve is the reason for this note. If you are completely satisfied with the performance of this product, fine. We're delighted. But if you are not, and if it did not in some way live up to your expectations, would you tell us about it? Just use the attached prepaid post card to jot down your comments. We value your patronage and want to make certain we continue to serve you to the best of our ability.

A WORD ON STYLE

Ask for Action

Every letter has a point. The reader should get that point as quickly and clearly as possible. He or she should not finish the letter thinking, "But what am I supposed to do? What does the writer want? Why was this letter written?"

Some letters, of course, are purely informative. A newsletter fits that description. The writer of a newsletter is not generally asking for any specific action. The reader is not expected to respond in writing. Letters of welcome, appreciation, and sympathy are other kinds that seldom call for a response.

Many of your letters, however, do require the reader to do something—to phone for an appointment, to attend a conference at a specified time and place, to fill out a form, to write a letter in return. With these letters, you should make your purpose absolutely clear. You should ask for action. You should specify what the required action is and when it needs to be taken.

This is commonsense advice, but you have most surely received letters yourself that failed to heed it, letters that left you wondering what your next move was supposed to be. Don't leave your own readers in the dark. When you expect action, ask for it.

Chapter 5

PRODUCTIVE LETTERS AND MEMOS ON TECHNICAL PROCEDURES AND POLICIES

Only the imagination can limit the approaches used in letters and memos on technical subjects. The wide variety presented in this book deal with many of the most common situations you face on a daily basis. Yet the utility value is enhanced by a fact referred to earlier: many of these letters fit neatly into more than one category. Depending on the situation, and the particular objective you have in mind, it is also possible to combine parts of different letters to provide one that is best suited to your needs.

Particularly when writing on technical matters it's often a good idea to first ask yourself two questions: "What information, specifically, do you want to convey?" and "How can it be conveyed most clearly?" Clear, firm answers to both questions will help reduce the necessity to include that popular last sentence: "If you have any questions, please let me know."

● *SECURING ACTION BASED ON DESIGN CHANGES*

General Guidelines

Changes in design, whether they take place in the preliminary stages or after production begins, set off a chain reaction that takes many forms. Letter 5-1 is one example. In addition to the revised design specifications themselves, letters to suppliers may be necessary to inform them of the need for new or modified parts resulting from the design changes. In some cases, this might lead to contractual problems. If a contract is in effect that binds you to the purchase of specific components for an existing design, it would be well to secure legal advice before sending such a letter. When the situation does not present a potential problem of this type, it may be dealt with in a less formal way. If it is strictly an internal matter, a memo to appropriate staff members may be sufficient. This is covered in one of the "Alternate Approaches" that follow.

Letter 5-1

ROYAL MANUFACTURING, INC.
1240 Meudel Boulevard
San Cristo, California 00000

May 5, 19—

Mr. George Smith
Kenyon Manufacturing Co.
7942 Distant Rd.
Elkton, Minnesota 00000

Dear Mr. Smith:

This confirms our conversation on design changes for the RX 100.

If you can furnish parts exactly as described in the revised specification sheet you left with me, in the quantities indicated and on the dates mentioned in your letter of April 28, you have a deal. We will need: (1) 4,000 of the A4719 silicon wafers on or before June 15, 19—, and 4,000 more on or before August 1, 19—, and (2) 10,000 of the X247B PC boards on or before June 1, 19—.

All items must conform to specs indicated in earlier letters, copies of which are attached hereto, and random samplings must pass our standardized tests as described in my letter of February 23. Please include these points in the sales contract you send to me for signing. I look forward to hearing from you soon.

Yours very truly,

Stanley Cormack
Chief Engineer

Alternate Approaches

(a) As we discussed yesterday during our telephone conversation, a number of recent developments have led to changes in our approach toward digital radiology. The enclosed listing will highlight the modifications necessary, but we would like you to pay particular attention to three aspects: first, the need to reconstruct 2-D images from 1-D projections; second, our preference for compression techniques that are less prone to error, including clipping and bit truncation, run-length coding, and Huffman coding; and, third, the need for a multiple modality display station that permits a review of images using video monitors. In view of the comprehensive nature of these changes, we would like you to revise your original proposal and, if at all possible, submit this at least one week in advance of our next board meeting on April 20.

(b) Please review circled portions on the attached blueprint and their related footnotes. They will pinpoint the design changes that have become necessary due to cost and time factors. We did not indicate all the resulting peripheral modifications because we need your expertise in designating these alternative approaches. For example, the revised dimensions and composition of the plastic rods are indicated, but obviously we'll need a different kind of bonding adhesive. This is just one of the things we'd like your advice and recommendations on. After you've developed your suggestions, please give me a call. Following our discussion, I'd like you to draw up a list of "things to do" as a result of these changes. When we reach agreement, I'll arrange to have our staff follow through by taking appropriate action on each item.

(c)   This memo will inform you of design changes in product X100 that were approved at the staff meeting on March 4. The attached revised specifications will identify these changes, and, as you will note, the Production Department expects to complete related modifications in assembly equipment no later than March 30. Would you please make certain that all of your appropriate people are notified of this. The quality control and marketing staffs will also receive copies. To make certain we're all on the same wavelength, I would like you to send me any comments, questions, or suggestions you have on this. Your input is important. The objective here is to increase the product's success, and your recommendations will help ensure this.

● *COMMUNICATING NEED FOR BUDGETARY ADJUSTMENTS*

General Guidelines

When writing to staff members on matters pertaining to the budget, as illustrated by Memo 5-2, try to show the relationship between the information you're conveying and the technical equipment and/or objectives they are concerned with. All departmental objectives, whether engineering related or not, are determined to a great extent by the funds available, and the ways in which those funds are used. It is easy in a world of complex details to overlook the ways in which a realistic budget can contribute to progress. In general, you will want to be particularly careful about making your meaning clear. The memo itself should have an objective, and its purpose should be one with which staff members can identify. Explain this purpose clearly and succinctly. If you are making a request, show how the response will benefit staff members as well as the organization as a whole.

MEMO 5-2

April 24, 19—

TO: David Atwater

FR: Richard Jenson

Dave,

I know you have been trying, but upon reviewing the last three months we find ourselves running 2,000 units behind budget, and your short-range forecast shows no improvement in the next few months. For the first three months of this year, you are behind the earlier forecast by a substantial margin; therefore, your costs should also be lower.

If you will evaluate the production analysis report, you will see how the loss of volume has affected profits. In addition to the drop in volume, I believe you are already aware that we are not maintaining last year's average price level, let alone the price level projected for this year.

Comparing the first three months with the first three months of last year, you have spent $7,000 more for printing, $19,000 more for cutting tools, and $18,000 more for packaging—all of these represent higher expenses with lower volume.

The only area in which you have reduced costs is maintenance, and that to the tune of $930. This is the only production-related item in which any improvements have been realized.

I would like a detailed memo from you indicating the steps you have taken to

1. Bring production volume up to your budget forecast level.

2. Correct the depressed average unit price levels.

3. Decrease your printing costs, bringing them in line with last year's costs.

4. Decrease cutting tool costs to conform to budgetary requirements.

Please plan to meet with me at 10 A.M. on May 8 so we can go over your written responses in detail.

Richard

Alternate Approaches

(a)   During our March 12 meeting, we agreed on the departmental budget that is now in effect. I refer to the date because it was not until the end of May that the X400 amplifiers went on the market and proceeded to astonish us with sales far beyond our original expectations. If we are to capitalize fully on this development, we must immediately give thought (and action) to adjustments in the budget. A capital investment must be made in additional assembly equipment as well as an expansion of our man-power resources—especially in the production and quality control areas. The last thing I want to do is make arbitrary decisions on this. The first thing is to seek your recommendations on specific modifications that are neces-sary in the budget. Would you please indicate your responses to the ques-tions attached, and return this to me no later than Monday of next week.

(b)   Our staff meeting yesterday afternoon uncovered a number of prob-lems that require immediate attention. Corrective action is required if we are going to meet our production goals for this quarter, but this memo deals with an even more basic factor that takes priority over everything else we discussed. The progress we make on the items we covered depends, to a very great extent, on revision and approval of our current budget. This must precede the changes in equipment discussed and the reassignment of person-nel. We need to know, without question, how much we have to work with in each area. And we won't know that until we've determined just how much the corrective action is going to cost us. The attached inquiries are designed to produce specific responses to the items listed. We need your best thinking on this. Please let me have your reply on or before May 8.

(c)   When I spoke to you on June 28, at the start of our budget planning sessions, I told you we would make every effort to develop a budget we could all live with for the balance of the year. So far, we have succeeded. Yet that time may be drawing to a close because of design imperfections uncov-ered in the relay assemblies manufactured for Smithson Corporation. This, as you know, led to a recall that has proved costly in many respects. Aside from the corrective actin that has already been taken, we must now rethink the budget because of heavy costs incurred as a result of this recall. That is not to say we can no longer operate under the current budget restrictions. However, we must carefully review the current situation and, if necessary, make appropriate modifications. I have attached a copy of the budget as it pertains to your operation. Please return this with changes you think must be made and your reasons for suggesting these changes. I'll need your detailed response no later than October 16.

● *RESPONDING TO SUGGESTIONS ON PRODUCTION PROBLEMS*

General Guidelines

Letter 5-3 points out one of the three responses that are generally given to suggestions. The writer may, as in this case, indicate he or she will "consider" the ideas offered. Or the response may indicate acceptance—or rejection—of the suggestions. In either event, one assumes that appreciation will be expressed in most cases to the one offering the suggestions, and some clear indication given as to the way in which they will be acted upon. When your response indicates you will (or may) accept and apply the suggestions, consider carefully whether you should include a closing phrase such as, "I'll let you know the results." If the situation clearly warrants this action, then make certain you follow through on it.

Letter 5-3

ACME MANUFACTURING CORPORATION
886 Hudson Boulevard
Kent, Michigan 00000

April 25, 19—

Mr. James Mallory
ABC Manufacturing Co.
3040 Industrial Rd.
Elson, Maryland 00000

Dear Mr. Mallory:

Many thanks for your suggestions regarding fabrication problems at the Gainesville plant. You can rest assured I will give them careful consideration. Things are finally beginning to look brighter at Gainesville, however, and I may end up making very few changes at this time. At least, until I can get down there and spend at least a week or so with the managers. In the meantime, I wanted you to know that I appreciate your concern and

```
your willingness to help. I'll be glad to let you know what
develops.

Sincerely,


Robert Kern
Chief Engineer
```

Alternate Phrases

(a)   I feel as if I owe you some kind of special award, or at least a free lunch. After seriously considering your suggestions on the Carlton matter, I decided to take the action you recommended. It worked like a charm. The Carlton people were delighted with the results, and it now appears we came out of this situation unscathed, thanks to you. I thought you would like to know, and whenever you want to take me up on that luncheon offer, I'll fill you in on the details. In the meantime, Bill, many thanks for your help.

(b)   It's always good to hear from you, and your recent letter was especially appreciated because of your suggestions regarding the fork lifts. Not, I must admit, because we acted on the recommendations, but because they reflected the fact you're thinking about us, and want us to succeed. The reason we finally decided to rent instead of purchase was based principally on cost considerations. Renting enabled us to avoid a substantial capital investment at a time when we could least afford it. In addition, we will save on maintenance costs. Under this arrangement, if a machine breaks down all we need do is call the equipment company, and it sends a replacement unit within a matter of hours. So far, our experience indicates this will save a considerable amount of time and money during the current year. Nevertheless, from a long-term point of view, I believe you were right to recommend an outright purchase of the units.

(c)   It's difficult to express how much I appreciated your suggestions on ways to relieve the financial bind we're in. This problem has had a devastating effect on our production output. As soon as I finished your letter, I knew that you had supplied us with the key steps to a practical solution, and I immediately began to take those steps. We fired off letters to all our creditors. As you suggested, we stated the situation in very clear terms, requesting them to permit deferral of payments for 60 days. We had no difficulty here. They all agreed. Next, we offered a 10 percent discount in an end-of-the-year prepayment schedule for our own accounts. Our regular customers leaped at this, and we raised $150,000. We're now in the process of implementing the other steps you recommended. Production has already increased, and the improved cash flow will enable us to purchase additional lab equipment that has long been needed. Thank you, Howard. We're most grateful for your assistance.

● DEALING WITH QUESTIONABLE REPORT DATA

General Guidelines

Memo 5-4 illustrates a reasonable approach toward situations that, in some instances, may involve a reprimand. There is an infinite variety of ways in which to convey doubt or disagreement. One productive way is to open with a positive reference to something that is satisfactory, proceeding to the negative aspects in a calm, controlled way. An abusive approach arouses defensiveness, resentment, and resistance, causing the reader to become unreceptive to suggestions for improvement. Wherever possible, offer encouragement and support, so the corrective steps necessary will appear easier to take. You should also include a concise, straightforward statement that clearly identifies what, exactly, is wrong. Set definite goals for improvement, and wherever necessary, indicate specific dates when the needed improvements must be implemented.

Memo 5-4

April 24, 19—

TO: James Darwell
      Systems Manager

FR: John Elkins
      Plant Manager

Cintex plant reports sent to headquarters during the past six months have been on time and have raised no significant questions, with one notable exception: the data processing has steadily deteriorated. This has delayed receipts of data for the sales statistic reports. It is also delaying our cash receipts because customers are getting their invoices late. The chain reaction also affects report data on new product planning.

While some of the problems can be attributed to the headquarters' Data Center, the major problems are being created by the Cintex plant. I understand the problems at headquarters will be resolved soon, and I'm confident that the future support from the Data Center will meet our standards.

The problems being created by the Cintex plant appear to be in three categories: (1) late entries past the cutoff date, (2) month-end bunching of production and sales data, and (3) other problems such as illegible copies and invalid data.

The attached schedule coupled with the related data that are part of it covers only the month of February. Just taking the matter of receipts as an example, you will note that 22 percent of February's invoice volume was apparently billed on the last two days of the month. With 20 billings a day, the expected percentage of billings on the last two days should not have exceeded 10-12 percent. Additionally, over 3 percent of the month's invoices were received after the cutoff date. You will notice similar problems that affect quality control and production planning.

I'm going to monitor future closings more diligently. Each month we will publish a report on the prior month's closing to highlight problems and take corrective action as necessary.

Please give this your personal attention. We must resolve this problem, and we must do it quickly.

John

Alternate Approaches

(a) I have usually been impressed with the comprehensive nature of your monthly reports. They have been helpful on many occasions. This makes it regrettable that I must question you about a serious omission in your reports for May and June. I have just received notification from the Fire Insurance Underwriters that our insurance on Plant 3 will be increased by 40 percent for the coming policy year. Upon inquiry, I was told that unsafe and even illegal conditions have been allowed to exist in your department and that this is the cause of the increase. I double-checked this and discovered you've had two citations from your area's fire marshall in recent months. Despite the fact our standardized monthly report forms contain a section relative to such matters, you failed to mention this situation in your May and June reports. I would like to meet with you in my office Thursday at 10 A.M. The fire marshall will join us. Please bring a written list of your recommendations on how we can correct this situation quickly.

(b)   Your status report for last month reflected improvement in most of the areas we discussed earlier. The bad news is that there was a discrepancy in the small motors physical inventory of November 20 compared to our book inventory. This amounted to approximately $39,000 (975 units). On a percentage basis, this added up to .8 percent. The division average is .2 percent, making our variance four times the division average. The effect of this variance is being explored, and I'll discuss this with you at our next meeting. However, in the future, it will be mandatory that we take steps to get this variance in line with the division average. Please go over this carefully with your staff. I would like a separate report from you, no later than March 15, on the procedural changes you will make and when you expect to put them into effect.

(c)   I know you have been trying hard to correct the manufacturing problems we discussed, but the Production Department report for May reflects insufficient progress. We now find ourselves running 20M units behind budget, and your short-range forecast shows only modest improvement expected in the next few months. If you will reevaluate the attached profit analysis study, you will see how this loss in volume has substantially lowered profits. I would like to know by May 7 what steps you have taken to accomplish the following: bring production volume up to the projected level indicated on the budget. Decrease unit manufacturing costs to the extent they are in line with 19— levels. Complete the redesign of our cutting tool operation, and finalize procedures for including product X-110 on the assembly line no later than July 1. This is no small assignment, Bill, and I want both of us to be constantly aware of its importance. I believe you can achieve these goals. Be assured that if you have questions, or need additional assistance, I'll be glad to help in any way that I can.

● *REQUESTING ADDITIONAL INFORMATION FROM STAFF MEMBERS*

General Guidelines

Memo 5-5 illustrates an orderly sequence to follow when requesting supplementary data on technical matters. In many instances, you will want to state initially the reason for the request. Once the reason is made clear, you can proceed to a specific listing of the data you need. The recipient of your memo may need to consult others to obtain the information. In this event, consider whether it would be advisable to send copies to those who would assist in providing the information. If you are not certain as to who they might be, you may wish to raise this point with the person you are writing to by suggesting copies be made and distributed to appropriate staff members. If the data you're requesting are complex, and will obviously take a good deal of time to compile, state your understanding of this. Wherever possible, allow the amount of time needed. If it is imperative that the information be received in what might be considered an unreasonable time period, give a compelling reason for this.

Memo 5-5

March 21, 19—

TO: Staff Members

FR: Donald Blackwell

We have been requested to provide the headquarters office with an updated physical inventory of lab equipment acquired last September. This updating is necessary due to the new fiscal year closing.

Please prepare your report in triplicate and attach a copy of the original inventory memorandum to each copy of your report. All three copies should be sent directly to my attention.

Please be certain that the cutoff is as of the end of business on September 30 and that all new equipment and supplies received since the last full inventory are properly recorded.

Alternate Approaches

(a)   I'm sure you can appreciate the need for our office to maintain up-to-date records on manufacturing equipment used in Plant #5. This actually benefits both of us because, as you know, when our records coincide it simplifies both the purchasing and maintenance functions. Recent discrepancies in our records cause me to request your help in updating our current list. We need your assistance, and any suggestions you have, regarding items that should be added or deleted from the current inventory record I have attached. We are now in the process of coordinating this information for all plants. In order for your responses to be incorporated in the overall corporate listing, I'll need your input no later than March 12. If you need to, please feel free to make copies of this memo for your appropriate supervisors.

(b)   Please note attached correspondence from the Environmental Control people. Our records indicate you took samples of the Worcester plant's smokestack emissions just recently. Do you have results of the tests yet? Our corporate office would like to be brought up to date on the requirements our emissions must meet now, and five years from now. We need these data so capital budgeting can be planned if additional filtering or other forms of treatment will be required. Please provide this information no later than May 1. In addition, we'll need your detailed response to the attached letter from Walter Reed. When you have this information, please call me. I'd like to discuss it with you before you send your written reply. Do not write to Mr. Reed. We'll do that after we receive the requested data from you.

(c)   I have looked over the proposed agreement for the sale of your heavy equipment. It meets all the usual requirements, but I do have a few questions and also want to go over the payment schedule with you. The die cutters are really not that old, and while you have applied the standard markup price, I think we could probably do better than this. Our feeling is just the reverse regarding the stamping presses. They are almost antiques at this stage, and the suggested price may be unrealistic. So far as the payment schedule is concerned, let's try to reduce the total time period from nine months to six. I doubt this would work any real hardship on the Abco people, and it would certainly help our cash position during the first half of next year. Please let me have your comments on each of these points no later than Friday of next week.

● *ACHIEVING SUPPORT FOR TECHNICAL CHANGES*

General Guidelines

As in the case of Memo 5-6, the writer should first indicate if he or she is in favor of a particular action before asking someone else to support it.

Next, it is incumbent on the writer to explain the basis for his or her affirmative position, and when necessary, the reason he or she is asking for support. Brevity is also important. If the request is prefaced by a lengthy recital of all the minute details and surrounding circumstances, this invites a negative reaction. Why, the reader may ask, should I support what appears to be a very complicated situation that the writer cannot express in simple, direct terms? It may also be desirable to make mention of the possible outcome if the reader does (or does not) lend his or her support. Anyone receiving a request of this type would naturally wonder what difference his support, or lack of it, will make.

Memo 5-6

March 21, 19—

TO: Fred Newton

FR: Al Meyers

Attached is Alex Smith's proposal to simplify our die usage calculations and make them more accurate. I feel that Alex did a fine job in his efforts to improve the procedure.

If you agree with his recommendations, Alex will be available to run a short training seminar on the use of the revised system.

Let me know how you'd like to proceed on this.

Alternate Approaches

(a)   At our last supervisory meeting I made a strong recommendation that we switch our entire data processing function to the new SW400 model. Because of this, I read with great interest *The Wall Street Journal* article that described the economies in time and labor you expect to achieve with this new and very versatile computer system. The article could not have come to my attention at a better time. It prompted a thought I'd like very much to share with you. First, a little background. We are a relatively small firm. Consequently we lack the experience you've had in this area, and, quite frankly, I am encountering some opposition from our board in making this change. On the basis of research we've conducted (prior to appearance of the article), I'm convinced we should follow through on a purchase of the SW400 system. Would you be wiling to serve as a consultant and advisor to us? I plan to call you next week so we can discuss this in more detail.

(b)   I have attached a copy of my Report to Management on the need for new compressor tanks in the Chicago plant. We did not keep records on the amount of time that went into this report, but I can assure you many hours were devoted to it by members of the engineering staff. When we put this much effort into something, we want that effort to be productive, and I think you can help ensure a favorable outcome. Based on your experience with a similar situation at the Bridgeport location, I hope you'll be able to join us at a management meeting on June 12 at 10 A.M. when our recommendation will be discussed in detail. I have attached a copy of the report itself, along with additional data that describe what we hope to accomplish at the June 12 meeting. Will you lend your support? A letter or memo outlining your comments and suggestions would really be helpful. Of course, if your schedule would permit you to join us on June 12, we would be delighted to have you. In that event, we would get together with you for a discussion prior to the meeting. Please give me a call after you've considered this.

(c)   I'd be the first to admit that the Quality Control Department has been of enormous help to us on many occasions. Usually, it's been a question of having defects or impractical procedures identified by your people and resulting corrections made by appropriate staff members or departments. Now, I'm going to ask your support on a recommendation that could be important to both of us. Specifically, we're about to suggest a design change in our Model 488. Initially, it appeared somewhat radical to me when I saw how extensive the modifications were, particularly in terms of time and cost considerations. After weeks of stringent tests, however, I'm now convinced the engineering staff has done a superb job. Please review the attached data carefully, along with the analysis of test results which were compiled by an outside firm. What I'd like you to do is this: provide a detailed overview of problems you've encountered previously with the Model 488. Then, relate this to the design changes we're recommending. Feel free to be critical wherever you think it's necessary. We want your frank, objective comments.

● *OVERCOMING TECHNICAL DATA
  DISTRIBUTION PROBLEMS*

General Guidelines

Memo 5-7 points out how the routing and distribution of various reports can contribute to, or subtract from, good communications. Sharing significant information with appropriate people is often not as simple as it might appear to be. The best laid plans can go awry, with a resulting loss of time and energy. The data itself may be expertly drawn, but if not conveyed to those people who can make the most effective use of it, there is little chance they will ever be put to fullest, most effective use. Countless problems are caused by what is often referred to as a "breakdown" in communications. In many instances, it is not the communication or data themselves that are at fault, but the procedure used to convey it. When writing to one individual it's usually a good idea to determine if others, perhaps on that person's staff, should also receive the data. If so, you may designate them and send additional copies, or request the recipient to do so.

Memo 5-7

March 28, 19—

TO: William Fife
     Quality Control
FR: Roger Johnson
     Systems Manager

This is a follow-up to our phone conversation of March 15 and your memo of March 26.

The problem referred to in your memo appears to have started earlier this year when some operations were discontinued at San Jose. At that time, we were requested to route several reports to San Mateo that previously had gone to San Jose. These instructional changes obviously did not result in proper routing of all reports affected.

We have issued report distribution instructions to eliminate the problem pointed out in your memo.

I would like to mention, however, that members of my group have found it extremely difficult to acquire and verify proper mailing instructions for reports. Suggestions such as the ones you initiated on March 15 are helpful in assuring correct distribution, and we appreciate your notification of the problem.

I encourage you to continue calling our attention to any problem of this nature. Please instruct Linda Arnette at Ext. 0000 to expedite the notification process.

I am sorry for the inconvenience caused by the erroneous mailing instructions.

Roger

Alternate Approaches

(a)   Attached are three requests for credits from clients who state they did not agree to pay extra for the molds used in making their aluminum castings. Their understanding was that the mold cost was included in the sales price. Upon investigation, I determined two things: first, they are absolutely right, and second, the reason this happened can be traced to the distribution of earlier instructions on how to handle a situation of this type. Please refer to the attached copy of procedures to follow when billing customers for castings. This was distributed to appropriate people in the Manufacturing and Production departments, but not to the Billing and Customer Relations departments. This memo should correct that oversight. Will you please distribute copies of the March 4 policy memo *and* this memo to Bill Winston and Howard Keely. They should see that their employees are fully informed of the correct procedure to follow.

(b)   I want to make certain each staff member is fully aware of the attached specifications. They relate to stainless steel brackets we are manufacturing for the Apex Corporation. Due to an oversight, we narrowly averted a serious problem that would have delayed delivery and, quite possibly, caused us to lose one of our most valuable customers. You will note that the specifications call for modifications in our standard length and thickness dimensions for these brackets. This was done in accordance with specific instructions we received from the Apex organization. However, if we had gone ahead with a first production run without informing Bill Snyder in the Quality Control Department, you know and I know what would have happened. He would have probably labeled the first brackets to come off the line defective, stopped production, and shut everything down until somebody told him what the facts were. Fortunately, he called me immediately. Please make absolutely certain that Bill Snyder's name is included on the "Special Job" routing list in the future.

(c)   It took all of us a long time to develop our staff procedures manual. A lot of thought and effort went into it. Yet the best procedures become meaningless if they are not followed regularly and properly. Please review the section on client relations beginning on page 12. If the instructions in this section had been followed, we would have been able to meet the deadline imposed on us by the Berkley Corporation. Instead, we almost lost this important account. Note on page 14 that our long-established procedure is to notify the chief engineer of production assignments that do not fall within the standard time allotments. We could have completed the Berkley switch assemblies within the time period they indicated *if* Howard Benson had been alerted to the situation. He was not, and as a result, Berkley had to wait an extra two days for the final shipment. Please make certain that your staff follows this procedure in the future.

● *ADVISING ON STAFF ORGANIZATIONAL MATTERS*

General Guidelines

Memo 5-8 tackles the sometimes difficult task of advising someone who may or may not have requested your guidance. Unless the giving of advice is mere balm for the giver's ego, it must be skillfully and diplomatically conveyed. This is particularly important when the advice is volunteered. First, the advisor must be careful to avoid putting himself or herself in an exalted position. It is also possible that the advice is diametrically opposite to the course of action preferred by the recipient. In this event, a negative reaction is almost guaranteed if the reasoning behind the advice is not sufficiently clear, and convincing. A tactful, diplomatic approach will also help, and wherever possible, one should preface the advice with a positive, supportive reference of some kind. This could have to do, for example, with satisfactory results obtained when the advice was applied to another, somewhat similar situation.

Memo 5-8

April 25, 19—

TO: John Rainer

FR: Walter Moore

John,

I am impressed with the detailed nature of your organization plans for the new quality control unit. I can't see anything you have overlooked, but of course setting up a new department is not my specialty.

I do have one thought you might want to consider. I noticed the salaries you projected for various technicians and supervisors were at the bottom of the scale. I assume this means you plan to take on inexperienced people and train them. Personally, I wish you'd think twice about this. I have known several managers who went this route, and they find it entailed innumerable headaches.

You may "save" on actual salaries, but you have to consider the length of time that such help will be nonproductive—sometimes even obstructive. There is also the cost of your time and of other managers that is taken up in training these people. In the end you may find you have spent more in this way than you would have spend on salaries for skilled, experienced help. (It also seems probable that at least one or more of the trainees will turn out to be duds.)

There is another angle to consider. That is, when you start this operation, you want to get it under way as soon as possible and begin producing. Please give this further thought. I feel you will have enough to do in the beginning without taking on additional manpower problems.

As indicated earlier, I admire your detailed proposals for the organization of this department. In every other respect, the overall plan seems a solid, viable projection of what needs to be done. Please continue to keep me posted, and also let me know what you decide to do regarding the level of manpower skills you will be looking for.

Walter Moore
Director, Engineering Services

Alternate Approaches

(a)   The realignment of the design staff has resulted in many of the improvements we discussed at the last management committee meeting. I'm sure you are gratified by this, and you have every right to be. Many of the suggestions for these changes came from you. Now, I'd like you to give some additional thought to one aspect that we may have overlooked earlier. This has to do with the reporting structure as it relates to corporate headquarters. To expedite the transmission of weekly and monthly reports to the central office I suggest you take two additional steps. Specifically, the reports in question should be routed to the chief engineer first. You should not be required to wait indefinitely for any response he may wish to make. My advice would be to send the reports to Bill Gabacia each Monday. Then, if he does not respond with questions or comments by Wednesday, send copies to Thomas Ridley at the headquarters office.

(b)   I was rather surprised to learn that you plan to go ahead on the purchase of a new power press for the Worcester plant, basing your action on advice from only one source. From an organizational standpoint, and looking at this objectively, the decision could come back to haunt you. My suggestion would be that you hold off until you've succeeded in making the "chain of command" work for you on this. Under the present circumstances your action could be viewed as arbitrary, if not impetuous. I'm aware of the need to increase productivity at Worcester, but it should not be brought about at the expense of other approaches that could be more cost effective. Please reconsider this. At the very least, I'd suggest you bring the purchasing manager into the picture, along with Howard Gainor, who is one of our most reliable outside consultants. You did not ask for my advice on this, Bill. I'm aware of that, but I offer these comments because I want you to achieve the objectives you've set for yourself at Worcester.

(c)   I'm flattered that you think I could advise you on your plans to reorganize the Production and Research departments. This is a monumental undertaking, and I'm glad to see you're approaching it in a careful, methodical way. First, I do not believe my experience is of the type that would be sufficiently helpful to you. That is not to say I don't want to help, because I do. And perhaps the best response I could offer would be to advise you on the selection of an expert in this area, one whose credentials are far above average and one you can rely on. To my knowledge there is only one individual in this immediate area who has a proven track record in staff organizational matters. I refer to John Hood of Hood, Blakely and Benson. If you wish, I'll be glad to pave the way for your discussion with him by setting up a lunch for the three of us. A brochure about his consulting service is attached. Give me a call after you've read this.

● *OFFERING SUGGESTIONS AND COUNSEL ON DESIGN SPECS*

General Guidelines

Whether offering or requesting guidance on design specs, Letter 5-9 points out the need to be specific and candid in your remarks. It is not enough simply to state your suggestions or ideas. A certain amount of embellishment is required, but not to the extent it clouds the issue. When offering suggestions it sometimes becomes necessary to point out that your motive is not entirely altruistic. In other words, you have a vested interest in the outcome. There is no reason to sidestep this if indeed it is a factor. It could also lend weight to the particular course of action you're suggesting. In any event, the letter or memo should reflect your role in the matter as that of an ally, not an adversary.

Letter 5-9

KOPPEL MANUFACTURING, INC.
810 Seagram Lane
Newport, California 00000

April 25, 19—

Mr. Robert Brown
3112 Home Street
Suburbanville, Montana 00000

Dear Bob:

I was interested to hear of your plans to use components listed in your spec sheet for the Tri-X product. I would urge you to talk to a number of people in the field before you take the plunge on this. I have the greatest respect for your engineering

abilities, but this particular area is probably more complicated
than you might imagine. You could avoid a great many headaches
through a preliminary discussion with someone more knowledgeable
in this area than both of us. If you plan to be in Chicago any
time in the near future, I'd like to have you go over your plans
with my colleague Richard Roe. He will be able to give you reliable guidance before you make any final decisions. Having known
Richard for several years, I'm sure he will be glad to help in
anyway he can.

Let me know a little ahead of time when you are going to be here,
and I'll make an appointment for the three of us to have lunch.
You can take it from there.

With best regards,

Alternate Approaches

   (a)   The preliminary construction plans for the Potomac project represent an excellent beginning. For the most part, I like the designs. There is
one aspect, however, I'd like you to investigate in much greater depth. The
cost analysis we made concerning the prefab approach just doesn't seem to
be as detailed and convincing as we'd like it to be. The designs we're using
appear ideally suited to this type of construction, and it seems unlikely,
with the present approach, we'll be able to capitalize on this to the fullest
extent possible. Please contact the Blackwell organization. Tell them we
need much more than a surface treatment of the subject. I'm sure you
agree that the time to resolve this question is now, not after we sign the
contract.
   (b)   Although I was opposed to it in the beginning, I'm glad you went
ahead with a prototype of the RL40 remote control unit. The production
costs turned out to be higher than we anticipated, but the eventual result
will save us far more than it cost us, since we know now, beyond any doubt,
that our design specs fall short of the mark. One of our prime objectives was
(and still is) to control devices up to 65 feet from this unit. The Testing Lab
now confirms that maximum range with the prototype is 40 feet. The
receiver unit should be rated at 15-amp resistive. The sample unit shorted
out at 10 amps. Please carefully review the attached detailed report from
the lab. Clearly, we must go back to the drawing board on this. After you go
over the report and discuss this with your staff, please give me a call. I'd like

you to pinpoint the changes necessary to bring this unit up to the original specifications approved by the director of engineering.

(c)   You have caught me at a bad time with your request for help. We're working against a deadline on the Brookhaven project, and two of our most experienced engineers are out ill. I do want to help, even though I can't get down there to spend some time with you on the problem. After going over the design specs, I have one suggestion you might want to consider. The assembly-line problems do not all stem from one source, and I suspect the major difficulty is caused by improper lubrication. When you redesigned the roller layout, it probably would have been a good idea to investigate a different method of lubrication. That leads to my suggestion. About two months ago we began to use a lubricating gel that does not have an oil base. Since any oily residue attracts dust, it's possible that this is contributing to the breakdowns you described. I have enclosed information about this gel. We use it for the conveyer mechanism in the Burlington plant and for reducing friction in our slide assemblies. A sample is enclosed.

I'll be interested in hearing from you after you've had a chance to try it out.

- ### *SECURING TECHNICAL INFORMATION FROM OUTSIDE CONTACTS*

General Guidelines

Letter 5-10 can be useful when seeking assistance from individuals outside your organization. Adaptations of this letter might be used for a variety of purposes. The goal might be to secure suggestions on product development, new uses for an existing product, time-saving methods in a manufacturing process, or improvements in other procedural processes. In the sample letter it is obvious the individual is seeking help. There is little point in disguising this fact. The writer also suggests this could develop into a mutually beneficial relationship. Again, it becomes clear that requests are more likely to be granted when there is the likelihood of a quid pro quo arrangement. Letters or memos of this type should not only make the request itself clear, but should give a reasonable amount of background so as to put the request in perspective for the reader.

Letter 5-10

PALMERTON CORPORATION
802 Brenner Avenue
Gibson, Pennsylvania 00000

April 24, 19—

Mr. Henry Anderson
District Manager
Large Manufacturing Co.
Industrial City, Massachusetts 00000

Dear Mr. Anderson:

I met you last year when I was with Jones & Co., and we purchased a wide range of PC boards from your organization. I have
now changed companies as you can see from the letterhead and
have relocated in nearby Gibson. It is a good location, but
current business conditions are making it rather difficult.
Also, some of the assignments I'm working on are relatively new
to me.

It occurred to me recently that you might be able to provide
some assistance with a few of the products we're working on—
along with perhaps a few introductions to other people who could
be helpful to us in the design of these products. Since the products both of us are involved with do not compete, but complement
each other, we might work out a system for a continuing relationship that would be beneficial to your organization as well as
mine.

In any event, I'd like to discuss this with you in more detail—
at your convenience. How about letting me take you to lunch
sometime next week? Would Thursday be a good day for you? You
name the place and time, I'll arrange to meet you wherever you
say.

Sincerely,

Jim Jones

Alternate Approaches

(a)   Several members of our association have remarked favorably about the excellent meetings we've had so far this year, particularly the ones held in May and June. The presentations made by both guest speakers were incisive, interesting, and exceptionally helpful. This prompts a request that I hope you will be able to fulfill. Actually, it could be applied to other meetings whenever they are as productive as these two were. Specifically, we'd like to have Dr. Wainwright and Thomas Hartley provide written transcripts of their remarks, for distribution to the membership. If this is done, I intend to make additional copies for distribution to our entire engineering staff. To defray any cost the association might have to bear in doing this, I'm certain the membership would make a contribution to offset this cost. I know that I would be glad to do so. Would you please put this on the agenda for our next meeting?

(b)   In my earlier letter of July 12 you were requested to provide research data on radar units comparable to the X400 models. The additional assignment that follows results from a directive that corporate headquarters sent to me this morning. A copy is attached. As you will note, it lists supplementary details which should be added to the July 12 request. With reference to the overall assignment, you are now requested to cover these additional points in your final report. Howard Snyder, one of our systems engineers, has some data that could be helpful to you. Please contact him directly about this. I believe the memo from headquarters was prompted by conflicting information received from other sources. Your report should clear up and resolve these differing points of view, and, as agreed earlier, we'd appreciate having this no later than August 15. If you anticipate a delay because of the additional data requested, please let me know as soon as possible.

(c)   This will confirm the request relayed to you during our telephone conversation yesterday afternoon. The general manager will require a physical inventory of all laboratory equipment on loan to outside sources, and we will need your response no later than September 30. It is essential that this deadline be met because of our fiscal year closing on December 31. Please observe the procedures described on the enclosed pages excerpted from our staff manual. The report should be prepared in triplicate and an inventory certificate attached to each copy. When you have completed this send all three copies directly to my attention. Please be certain the cutoff is as of the end of business on September 30. We must be certain that all receipts and shipments or deliveries of equipment up to the time of taking the physical count are properly recorded.

● *RESPONDING TO REQUESTS FOR TECHNICAL ASSISTANCE*

General Guidelines

Letter 5-11 is made simpler by the writer's willingness to render the assistance requested. When you feel you must decline a request for help, and yet do not wish to offer a detailed reason, there are a number of approaches that will serve your purpose. You might suggest that the answer to the problem probably lies within the requester's own organization, and not being fully aware of factors affecting that organization, you do not feel qualified to suggest a solution. You should also bear in mind that acceding to a request obligates you to assume a portion of the responsibility for the results, provided your suggestion is put into effect. To offset this, you might consider a modified "disclaimer." For example, "I have no way of knowing whether this idea will help you. I only know that it worked for us when we were faced with a similar problem."

Letter 5-11

AURORA CEMENT CORPORATION
112 Tasley Avenue
Accomac, Maryland 00000

April 24, 19—

Mr. Richard Chambers
Ames Contractors, Inc.
912 Carson Street
Hampton, Ohio 00000

Dear Dick:

Your inquiry regarding assistance was well phrased and you can be sure the problem is not an uncommon one. I think cement wall

construction might be the answer to some of the cost problems
you are running into on the Brookfield project. Since we have
had considerable experience with preformed concrete, I'd be glad
to spend some time with you and discuss the possibilities this
approach offers if you are interested. I might even be able to
help you out with supervision in the beginning, or lend you one
of our men, if the process appeals to you. I can't forget how
helpful you were to me years ago when I was new in this area. Let
me know if you want to set up a meeting so we can explore this
further.

Sincerely,

Alternate Approaches

(a)   Thank you for expressing confidence in my ability to advise you on problems with hydraulic equipment in the Acme plant. I wish I were equally confident of my skills in this area, but I am not. You may have felt this way because you're aware of similar equipment we have in our Baltimore plant. It is true that we went through a period of excessive maintenance costs, but the solutions eventually arrived at were primarily developed by the manufacturer of the equipment. During that entire period I maintained a "hands-off" attitude because I felt the manufacturer not only had the necessary skills, but also a clear-cut obligation to fix what needed fixing. If I have any suggestion at all, it would be for you to insist (more strongly, if need be) that the manufacturer and/or supplier of your hydraulic Travelifts provide you with cost-effective solutions to the problems you described.

(b)   I can appreciate the difficulty you've been having with the cable assemblies described in your letter. In offering this suggestion, though, I'm reminded of the old saying that free advice is usually worth just about as much as you pay for it. This is just another way of saying that I am not at all sure my idea will solve your problem. Nevertheless, I can recall having similar difficulties a year or so ago, and we eventually discovered that the matching transformers we were using were defective. This seemed like a one-in-a-million possibility, and we didn't check it out until we had exhausted every other possible fault. So, the only suggestion I can think of at this point would be for you to have your staff check out the transformers. I assume the ohm rating for the antenna you're using is appropriate for a 75-ohm coax downlead. The enclosed schematic shows the circuitry we're using, and if you're able to adapt this to your situation, feel free to do so.

(c) As you say, there are two ways that you can go in solving the problem, but I would not want to take the responsibility of advising you on this. For one thing, I just do not know enough about the application of robotics to the situation you described. It is true that we entered this field last year, but our assembly team makes only limited use of robotics, and we just don't have enough data yet to base any firm judgments on. The only solution I'd feel comfortable with at this point would be to raise the possibility of having Tom Crowley call you to see if he could lend a hand. "Lend" is probably the wrong word because Tom was the expert who guided us in our installation, and he does this for a living. I've enclosed some data about his consulting organization, and you will note that their fee schedule is on the high side. This, however, is relative, and if he is able to do the job you want done, the eventual savings could outweigh the initial investment. Please let me know if you'd like me to have Tom call you for an appointment.

● *COMMENTING ON TECHNICAL REPORT DATA*

General Guidelines

Observations on technical report data, as in the case of Memo 5-12, should pinpoint the specific details you're concerned with. While letters and memos of this type may be critical or complimentary (and sometimes both), it is usually desirable to make your comments brief and to the point. An extensive coverage should be separated into items that can be listed and numbered. This contributes an orderly progression to your observations, while enabling the reader to be specific in any response that is called for. When there is any question in the writer's mind as to his grasp of the technical matters being reviewed, this should be stated. To do so represents a wiser course than, for example, criticizing a particular detail without fully understanding all the factors involved. Your concern can be expressed in the form of a question rather than making an allegation that is not clearly supported by the facts.

Memo 5-12

April 23, 19—

TO: Richard Cromwell

FR: Paul Dickson

The discrepancy in our small motors physical inventory of November 20, compared to our book inventory, amounted to approximately $39,000 (975 units). On a percentage basis this amounted to .8 percent. The Division average is .2 percent, making our variance four times the Division average.

In this particular case, no one is objecting to the favorable effect of the variance, but in the future it will be mandatory that we take steps to get this variance in line with the Division average.

After you have had a chance to review this, please send me a note by May 15 on the procedural changes you will make.

Paul

Alternate Approaches

(a) Your report of May 12 appears to be the result of considerable effort, and I commend you for this. It raises a number of questions regarding procedures in our maintenance shops, and I would like to see one of the recommendations supported by an analysis of practices followed in the industry, specifically, the production and record-keeping aspects. Many trade associations in our field collect data relative to the procedures you covered, and I think we need to compile a wider range of practical and usable information on this. The Birmingham chapter of the Commerce and Industry Association might be helpful. The research department at the headquarters office may have useful information too, and Al Jordan should be able to direct you to sources of useful data. After you've completed the assignment, I'd appreciate receiving a summary, along with a list of any changes this might exert on recommendations made in your May 12 report.

(b)   It appears that the cost analysis performed by the technical staff has uncovered the tip of a very large iceberg. We're going to have to go into this much more carefully than originally contemplated. In this respect, the analysis has been helpful. At least we now know a little more about the dimensions of the problem. What I'd like you to do, first, is call a meeting of supervisors from the maintenance and production staffs. I would like to attend that meeting, and will be available any day next week except Thursday. We should also bring Bill Caldwell into this picture. You may not have met him before, but he is the principal consultant who worked on improving methods and systems in the departments covered by your analysis. Please immediately draw up an agenda for the meeting next week and let me see it before copies are distributed. After we agree on the points to be covered, I'll call Bill Caldwell and arrange to have him join us at the meeting.

(c)   Your report on the defective compressors should be helpful to all the departments concerned—helpful, that is, from the standpoint of identifying the primary causes. This should lead us to corrective measures that need to be taken. At the same time, though, I think we have to probe more deeply into the effect all this will have on our current and projected costs. From a budgetary standpoint, it is imperative that we do this, and do it quickly. There are also costs involved that would not necessarily be reflected by the budget. For example, what have these rejects cost us in terms of customer relations? Were any accounts, large or small, lost as a result? If so, what should we do to make amends? Should our guarantee for the replacement equipment be extended? If so, for how long? These are just a few of the questions we need to consider. Please develop an addendum to your June 4 report that covers these additional points, along with the specific ways in which you plan to resolve each one. When this is ready (no later than June 30), please call me.

● *CLARIFYING SPECIFICATIONS RELATIVE TO BID SUBMISSION*

General Guidelines

Letter 5-13 reflects the need occasionally to expand on design specifications after a particular bid has been accepted. Few letters or memos require greater attention to detail than do those dealing with the submission of a bid. To the extent communications of this type are succinct and convincing, the likelihood of completing the assignment is increased. This section deals not with the bid itself, but with related communications that can help clarify, strengthen, and supplement the formal proposal. Brevity is desirable, but not always possible. The sample approaches described here are presented, for the most part, in summary form. When lengthy presentations are absolutely necessary, using wider margins, double-spacing, and indented, numbered paragraphs will help give the document a less formidable appearance.

Letter 5-13

HILLSDALE CONSTRUCTION CO., INC.
422 Kecoughtan Road
Langley, Vermont 00000

April 22, 19—

Mr. Thomas Benson
Benson & Benson, Architects
808 Forest Street
Hopewell, New Jersey 00000

Dear Tom:

Your proposal passed with flying colors, and Fairfield has given us the Hotel Plaza job. Construction should start next fall, but we'll need your talents in the meantime for a few other bids we're working on.

There's one segment of your presentation on the Hotel Plaza design that I'd like you to expand on. Specifically, the Town Council has asked for an expanded review of that portion of your proposal dealing with zoning regulations and the ways in which your plan relates to each one. This can be in summary form. While they are seeking an overview of this aspect, they appear to be especially interested in safety factors that will be built into the project.

As I told you, our specifications needed that extra polish provided by the precise way in which you addressed their other major

concerns. Please let me have your remarks on zoning regulations
no later than May 15.

Many thanks, Tom.

Richard Moore

Alternate Approaches

(a)   The signed contract from Easton Mfg. arrived today, and it should
help raise our sales volume for this quarter by a substantial margin. You not
only managed to attract an important new client, but one that has eluded
our best efforts over the past few years. Well done, indeed. Easton added a
codicil to the contract that I think we can live with. It goes beyond the
terms specified in our original bid, but it does not materially alter the
situation. Specifically, and as you can see from the attached copy, Easton
wants a slight modification in the reinforced concrete we intend to use. I
have checked with Bill Kaufman, who will be chief engineer on this project,
and he assures me the cost differential will be under $1,500. So we plan to
go along with them on this, provided you agree. After going over the
attached material, please give me a call. We'd like to send copies of the
countersigned contract to Easton as soon as possible.

(b)   The fact that we lost out on a very important bid to the Allison Group
resulted from errors in our purchasing procedures. This was a real disappointment to all of us, and it represents a situation that we should try to learn
from. After a good deal of investigation, we've determined the primary
cause. The overall responsibility for control of major raw materials, process
chemicals, maintenance, and capital equipment is vested in the headquarters
Purchasing Department. We are now committed to a policy of buying materials and services at the division, mill, plant, or office closest to the point of
ultimate use, commensurate with sound purchasing practice. In a sense, this
is a system of decentralized buying with centralized control. We lost the
Allison account because this procedure was not followed. From now on,
purchasing by headquarters will be done only for those divisions that do not
have a purchasing unit or where headquarters is able to secure a lower price.

(c)   On June 18 we were invited by the Raynor Corporation to bid on the
manufacture and installation of 28 AC motors for its Bloomington plant.
The technical staff spent hours and hours in the development of this bid and
obviously did a good job. We received confirmation this morning that our
bid has been accepted. We could all be even more pleased with this situation

if one element had not been missing when we sent the proposal to Raynor. As matters now stand, once we deliver and install the motors, that is the end of the line. However, if we had submitted a stronger proposal on follow-up maintenance at the same time we sent the primary bid, it could have helped ensure a continuing relationship with Raynor. Of course, it's not too late to do this, and we expect to send this out shortly. But when confronted with opportunities like this in the future, please make it a standard procedure to consider whether a more detailed maintenance agreement should be included with the bid.

● *SUGGESTING PRACTICAL IDEAS FOR STAFF BULLETINS*

General Guidelines

The idea of issuing periodical bulletins, as reflected by Memo 5-14, can be an effective means of regularly reminding staff members of policies or procedures that are particularly important. As suggested by the word itself, a "bulletin" is generally short and to the point. Its value, and the impact of the message, is diluted by trying to cover too many subjects. Once you've determined exactly what you want the bulletin to accomplish, it becomes easier to explain the purpose of the communication to those who receive it. A few additional guidelines would be to state the message briefly, clearly, and in a way designed to enlist support for the objective you're seeking. The manner of distribution depends not only on the message itself, but on the facilities you have at your disposal. Consider, for example, if you would be more likely to reach the appropriate people by sending copies directly to them instead of (or in addition to) posting it on a bulletin board.

Memo 5-14

March 21, 19—

TO: Martin Banks

FR: Kenneth Chappel

Because of the recent increase in our accident frequency
rate, headquarters has suggested we issue a monthly bulletin

to publicize our safety activities. This would be separate from
the monthly figures that show the number and type of accidents
reported. The monthly bulletin would contain reminders, suggestions, new equipment ideas, new safety equipment being made
available, etc., etc. I would like you to begin work immediately
in developing the framework for this bulletin, which should
probably be no longer than one or two 8 1/2″ × 11″ pages (double-
spaced) in length. This would be posted on all bulletin boards
and should contribute to a greater safety awareness among staff
members.

Please start on this as soon as possible so we can issue the
first bulletin by the end of next month.

Kenneth

Alternate Approaches

(a)   At a recent management meeting I asked our methods and systems
people to supply, on a regular basis, cost-saving ideas and suggestions
which could be circulated to supervisors in the Engineering Division.
Since representatives from this group are active in so many departments,
we might get a variety of worthwhile ideas. They have agreed to do this.
The plan is to issue occasional bulletins containing solutions or useful
approaches to problems encountered by other departments, with an eye
toward adapting some of these ideas to our own division. I have asked
Walter Benton to quarterback this activity. In the beginning we won't try
to maintain a regular schedule for these bulletins. Walter, will, however
try to issue at least one a month. Then, if it proves to be a productive idea,
we'll work out a procedure that involves sending them out on a regular
schedule.

(b)   In line with the policy of issuing occasional bulletins to our technical writers, here's another tip that might be helpful. H. W. Fowler, that
long-time arbiter of prose style, warned against what he called "elegant
variation." It bothered him that many writers seemed compelled to choose
synonyms rather than repeat the same word frequently. For example, a
committee would become a group and then a panel, and then a team—all
to avoid repeating the word "committee." The problem is that "elegant
variation" is sometimes more likely to confuse the reader. He or she may
assume that subtle distinctions are being hinted at, when none in fact exist.

The "committee" does not change its nature from sentence to sentence. It is the same old committee, but overzealous reference to a thesaurus may cloud that fact. Thus, it may be best to stick with your original (and most accurate) word, particularly if that word is vital to your reader's understanding of an entire memo or letter.

(c)   The shipping department recently sent a memo to managers on the use of zip codes. They requested that the information it contained be circulated to appropriate people within each department. One section of the memo dealt with the two-letter abbreviations for state names. This particular reference seems a good one to include in the staff bulletin we send out each month. Specifically, there are so many state names that begin with the letter "M" that mistakes are frequently made when using the two-letter abbreviations for these states. Here they are: Maine/ME, Maryland/MD, Massachusetts/MA, Michigan/MI, Minnesota/MN, Mississippi/MS, Missouri/MO, and Montana/MT.

While I have attached a copy of the shipping supervisor's full memo, I think the reference to the preceding state names is the only portion we need to include in our bulletin.

- **REVISING PROCEDURES TO ACCOMPLISH STAFF OBJECTIVES**

General Guidelines

Changes in procedures, as illustrated by Memo 5-15, should not only be described clearly, but the reason for the change should also be stated. It is not unusual for a procedural change to be instituted and then have it receive less than enthusiastic staff support because no one understands the change or disagrees with its purpose. Revision of an existing policy or procedure should seldom, if ever, be communicated orally. When the situation is such that verbal instructions are issued, they should quickly be followed up by written confirmation. If the writer is relaying a change that has been instigated by someone on a higher level, there should be no question as to whether the writer supports the action. Not to indicate support would be to invite the same response from those who receive the communication.

Memo 5-15

March 21, 1987

TO: Martin Woods

FR: Thomas Carter

In order to meet shipping schedules without increasing the personnel and equipment of our Shipping and Receiving departments,
we must restrict the receiving of raw materials to either the day
or swing shift. Our primary supplier prefers the swing shift, but
this may interfere with our secondary supplier with whom you deal
directly. It could also have a direct bearing on our quality control procedures.

Please investigate this from the secondary supplier's point of
view, and let me have your recommendations no later than March 30.

We must act on this soon.

Tom

Alternate Approaches

(a)   At the staff meeting on May 4, a question was raised concerning the
warranty for Model X100. It was a good question. Specifically, would the
excellent reliability track record established by this product justify an extension of the warranty period. During the intervening three months, a
study was made of this, and the results provide a convincing answer to the
question. There were three significant facts uncovered. One, the product
returns data indicate less than 1 percent were returned because they were
defective. Two, the quality control supervisor confirms that this product
has never presented a warranty problem during the 16 months it has been
in production. Three, our marketing people tell us the competition grants,
almost without exception, a three-month warranty period. Thus, an extension of our warranty to six months could help to stimulate sales. For the
reasons indicated, and as of October 1, our warranty for Model X100 will
be extended to six months.

(b)   Our Parts Department has suggested we revise the specifications form we use when purchasing supplies for the Waterford plant. Apparently we sometimes permit an outside vendor to handle transactions over the telephone without following up with written confirmation of the terms. Al Borden suggests, among other things, that we have a statement along the following lines and that this should appear at the top of the specifications form: "If you can furnish parts exactly as described on this specifications sheet, in the quantity and on the dates indicated below, please confirm this in the space indicated. All parts must conform to appropriate Milspecs, copies of which are attached hereto, and random samplings must pass our standardized tests as described on the enclosed form. By submitting quotes on the items that follow, it is understood that you are in agreement with the terms indicated."

(c)   Please carefully note the attached complaint letters received from customers who purchased our new optical scanner in the last two months. Clearly, we need a new packing procedure to prevent damage during shipment. The methods systems supervisor suggests we investigate foam-spray packaging, particularly with the X380 balance wheels, which seems to suffer the most frequent damage during shipment. I have asked Warren Berkley to go over this problem carefully with his staff and make recommendations. Warren has had some experience with the Stellar Age foam spray product. He says the material firms up instantly, and when used properly, the resulting package is just about as shockproof as any package could be. When used in quantity, the cost is not prohibitive, and certainly it would be less expensive than the problems caused by having irate customers. Please make certain that this packing procedure for the optical scanner is revised and put into effect at the earliest possible date.

A WORD ON STYLE

Use Abbreviations Carefully

Many organizations, departments, agencies, and programs are better known by their abbreviations than by their full names. A few examples are IEEE, AAASE, EIA, and AASA. You would not ordinarily use abbreviations of this type in a letter to a layperson.

But it is very easy to let familiar abbreviations slip unexplained into letters and memos. Surely, you think, my readers know what IEEE means. Do they? Perhaps. The test is this: if every single reader of what you write is likely to know what your abbreviations (or acronyms) stand for, there is no problem. If a few of them are left in the dark, however—if a few of them sit there, letter or memo in hand, groping for your meaning—then it might have been better to avoid those bunched capitals, or explain them in parentheses, or spell out the whole abbreviated name.

Chapter 6

COMMUNICATIONS INVOLVING COMPLIMENTS, COMMENDATIONS, AND CRITICISM

Whether you are involved with an individual engineering practice or work for a large corporation, it is essential to recognize the importance of good client and employee relations. One way to show your awareness of this is to express appreciation on those occasions when it is warranted. The common perception is that thank-you letters are easy to write, while those that convey criticism are much more difficult. If this is so, one wonders why more thank-you letters are not written. One reason is a natural tendency to gravitate toward problems that need to be resolved and letters and memos that will help resolve them. Yet an effective way to prevent many of these problems from ever happening is to convey appreciation, compliments, or commendations when they are clearly called for. Doing so reflects not only common courtesy, but good business sense.

This chapter covers a wide range of situations, with the emphasis placed on positive, rather than negative, approaches. We will cover some of the more common opportunities to maintain (and improve) professional relationships, as well as the dilemmas that result from such things as poor work performance by staff members. The letter or memo that best reflects a genuine spirit of friendship and consideration is a natural and somewhat informal note written to an individual. Critical situations involving reprimands will, of course, require a more carefully documented and formal approach. When conveying appreciation for the relationship you have with a valued client, avoid any obvious sales message. As pointed out earlier, and regardless of the nature of your remarks, make the message only as long as it needs to be. Appreciation, excessively conveyed, becomes maudlin. On the other side of the coin, going on and on about a staff member's shortcomings is counterproductive.

● *RESPONDING TO CLIENT'S LETTER COMMENDING A STAFF MEMBER*

General Guidelines

A letter of commendation represents an opportunity to not only thank the individual but to reciprocate the feeling of goodwill originally expressed by the writer. Letters of this type are all too rare. They should be viewed as another and very effective way for you to cement good client relations. More basically, any client who has taken the trouble to write a complimentary note regarding someone in your organization deserves an equally thoughtful response. Letter 6-1 or the "Alternate Approaches" will help you draft an appropriate response.

Letter 6-1

TRANS WORLD MANUFACTURING, INC.
380 Madison Avenue
Albany, New York 00000

March 7, 19—

Mr. Walter Nemeth
520 Biscayne Blvd.
Phoebus, Virginia 00000

Dear Mr. Nemeth:

Thank you very much for your kind letter of April 16 commending Ms Donna Jones for the technical assistance she provided to you. It is always gratifying to receive a letter such as yours, indicating that a member of our engineering staff has displayed the kind of courtesy and helpfulness so essential to any organization. You may be sure that Ms Jones will receive appropriate recognition and commendation from her immediate supervisors, as well as my own thanks and congratulations on a job well done.

We also appreciate your selection of equipment manufactured by our firm and look forward to being of additional service to you in the future. Needless to say, I hope you will find the services of all members of our staff to be satisfactory in every respect.

Thank you again for your courtesy and thoughtfulness in writing to me.

Sincerely,

William Bailey
President

Alternate Approaches

(a)   Your letter of April 16 made my day, and I am certain it will do the same for Donna Jones. As you said, most people delay writing letters of this type, but are quick to register a complaint when service is inadequate. You, obviously, are an exception, and we sincerely appreciate this. You will be glad to know we are aware of Donna Jones's courteous and efficient manner in dealing with clients and feel certain she will continue to progress in our organization. A copy of your letter will be sent to Donna, along with our own expression of appreciation for her exemplary job performance. This matter will also be made a permanent part of her personnel file. Thank you, Mr. Nemeth, for your thoughtfulness.

(b)   We do appreciate your gracious comments regarding the technical assistance rendered to you by Ms Donna Jones. Perhaps it's a bit immodest, but it appears that great minds do think alike. In other words, we agree completely with your evaluation of Donna's attitude toward clients and her skill in resolving highly technical problems. That is why we look forward to the day when we will be able to increase her level of responsibility in the Engineering Department. Letters such as yours will certainly help Donna in her progress toward a supervisory position. Your comments are also a reminder to us of how fortunate we are to have a staff member of Donna's caliber. Thank you very much, Mr. Nemeth. You may be certain that Ms Jones will receive a copy of your letter and appropriate recognition and commendation from the management at Trans World.

(c)   Our sincere thanks for your exceptionally thoughtful letter regarding a valued member of our technical staff, Ms Donna Jones. While we're always glad to receive appreciative notes from our clients, I was particularly pleased to read your comments regarding Donna. She is a relative new-comer to our organization, and it's gratifying to know she has progressed so rapidly in only a few months with our firm. I'm sure she will be delighted with your letter. We certainly are. A copy will be made part of her personnel file, along with a complimentary note from her supervisor. Your letter also gives me an opportunity to express our appreciation for the warm and cordial relationship our firms have enjoyed over the years. For my part, I feel privileged to be a part of this relationship, Mr. Nemeth. We look forward to being of additional service to you in the future.

● *DOCUMENTING OUTSTANDING PERFORMANCE BY A STAFF MEMBER*

General Guidelines

Expressing appreciation for a co-worker's skill is one way to ensure con-tinued excellence on the part of that particular staff member. In a prover-bial sense, it is bread cast upon the water. Very often, the goodwill and gratitude conveyed by the writer is returned in an equally positive form by the recipient. The sincerity of such a memo can be emphasized by its brevity. State the reason for your memo concisely. As in the case of Memo 6-2, make certain it becomes a part of the staff member's personnel file. If the individual is not a member of your staff, a copy should be sent to his or her immediate supervisor.

Memo 6-2

February 17, 19—

TO: James Dunnellon,
     Personnel Department

FR: James Benton,
     Chief Engineer

```
CC:  William Tavares,
      Quality Control
```

Please retain this memo in the personnel file for William
Tavares.

Every manager's dream is to have assistants who do not add to the
manager's work load, but do in fact reduce it. William Tavares is
that kind of assistant. He never comes to my office to discuss
problems without also bringing possible solutions—and more often
than not they are acceptable, well-thought-out alternatives.
William Tavares is an alert individual who frequently antici-
pates and heads off problems.

In the three years he has been with our organization, he has
earned the respect of his peers. I have never heard a complaint
from any staff member regarding his attitude or performance. His
approach to engineering problems, particularly with respect to
the element of quality control, is considerably above average.
This memo should become part of his permanent employee record.
In the meantime, Mr. Tavares may rest assured his exemplary performance is recognized . . . and appreciated.

Alternate Approaches

(a) You have heard me complain on occasion about defective products
that escaped the attention of our quality control people. This memo is a
distinct change of pace, and I am glad for the opportunity to send it to you.
It has to do with the superb job William Tavares performed during our
changeover to new assembly-line procedures. Mr. Tavares not only managed to identify a number of errors in the original design of the system, but
he contributed ingenious and effective solutions to problems that occurred
during its installation. I have thanked him personally, but as his department head, I'm sure you'll want to know how pleased we were with his
invaluable assistance. A copy of this memo is also being sent to the director
of human resources at the headquarters office.

(b) I know you will be pleased to hear that the X111 project was completed on time, and within the budgetary limitations established earlier.
You will probably agree that this achievement was largely the result of a
team effort. Nevertheless, teams are comprised of individuals, and I am
convinced you are the individual who helped immeasurably to spark this

particular team into such an outstanding effort. You can be proud of your contribution, Bill. Copies of this memo are being sent to the Engineering Management Committee and to Paul Atkins in the Personnel Department. Thank you for a fine job.

(c)   The successful design of a new product is generally attributed to sound engineering practices, effectively applied. While this is certainly true of our work on the new R12 hydraulic lifts, there was a particular contribution I'd like to draw to your attention. Specifically, the research work completed by Mary Endicott. The data she uncovered was extraordinary in the sense that it saved us an enormous amount of time during the original design stage. Not only did it help us stay on schedule, but it prevented expenses that would have exceeded the budgetary limits imposed by management. The fact Mary is not an engineer would astonish some people. Yet, in this instance, the entire engineering staff owes her a debt of gratitude. This memo expresses our sincere appreciation, and a separate copy is attached for inclusion in her personnel file.

● *INFORMING STAFF MEMBER OF PROMOTION*

General Guidelines

Informing an employee of his or her advancement represents not only good news for the recipient, but is a gratifying assignment for the writer as well. This is an opportunity to emphasize superior performance and ensure a productive relationship in the future. Letter 6-3 clearly identifies the new job title, and whenever necessary, you should also spell out the specific responsibilities involved. Convey your good wishes for the employee's success in the new assignment, and your support for their continuing efforts to advance within the organization.

Letter 6-3

AMERICAN VISCOSE CORPORATION
6333 East Corsair Street
Los Angeles, California 00000

March 28, 19—

Mr. Robert L. Daly
American Viscose Corporation
212 Amana Street
Waltham, Illinois 00000

Dear Bob:

It is with great pleasure that I inform you the Board of Directors has approved the recommendation that you be designated the new director of the research staff.

There isn't another person that I know who deserves this advancement more than you. Your efforts these past few years have contributed substantially to the success of our entire organization.

Congratulations, Bob. You have really earned this promotion.

With warmest regards,

Alternate Approaches

(a)   You may recall the conversation we had some time ago when we discussed your future with our firm. Your outstanding work was recognized, and you were told we hoped to reward your efforts in the not too distant future. I am happy to tell you the time has come when we can now express our appreciation in a more tangible way. Effective May 1 you will be the new product manager for the Electronics Group. This is an important assignment, and one you have earned by your consistently fine effort as technical advisor to this group. A detailed job description is attached to this memo, and I'm sure you will find it a most challenging assignment. Obviously, we are pleased with your performance, Bob. I hope you will be equally pleased with this new opportunity, and I extend my very best wishes for your continued success.

(b)   I am delighted to tell you that your promotion to computer operations manager has been approved. You have worked long and hard for this advancement, and it is reassuring to know that this outstanding effort has been recognized. While the problems and challenges confronting this department continue to multiply, I feel certain you are capable of achieving the objectives listed in the attached job description. Rest assured you will have my support. After you've had an opportunity to go over the responsibilities involved, let's get together to discuss the details. We'd like you to take over on June 15, so we should have our meeting within the next few days. Give me a call when you're ready, and once again, my very best wishes to you on this new assignment.

(c)   Congratulations. Your promotion to chief design engineer will become effective on June 12. On the basis of your superior performance as assistant manager, I have no doubt you will succeed in this new and more responsible position. Management concurs in this view, and we look forward to getting your thoughts and ideas regarding the operation of this important division of the company. A detailed overview of the department is enclosed, along with a description of the manager's responsibilities. On July 12, one month after you take over, I'd like to meet with you to get your preliminary thoughts on departmental objectives and ways in which you think they can be most effectively achieved. Every good wish to you on this new assignment, Tom.

● *CRITICIZING EMPLOYEE'S ATTENDANCE RECORD*

General Guidelines

In conveying criticism, every effort should be made to focus on the cause of the problem and its effect on the individual involved. A factual presentation is required, not an emotional harangue. A memo like 6-4 is a written record that can come back to plague the writer if it is not reasonable in tone and specific in its assertions. An ambiguous approach to the action required should also be avoided. There should be no doubt in the employee's mind that you have presented the facts clearly and that the situation requires the corrective action referred to in your memo.

Memo 6-4

March 31, 19—

TO: Harvey Wilson

FR: John Benchley

Harvey,

As I told you earlier today, your work has been excellent—but your absentee record is about to overshadow your work record. While I am sympathetic to the problem you referred to, there is just no way we can continue our relationship without a substantial improvement in your record of attendance. You must be aware of the difficulties imposed on your supervisor, co-workers, and scheduling operations when we cannot depend on your attendance.

This subject has been discussed several times previously. Now, your attendance must meet our requirements, or we will definitely have to consider termination. On May 1 I will review your attendance record with you.

Alternate Approaches

(a) This memo will confirm the key points covered in our discussion earlier today. It is evident that our talks on three previous occasions have not brought about the necessary improvement in your attendance record. Now, despite the fact I don't like ultimatums any more than you do, the situation forces me to take the following approach. If, at any point during the next three months, your attendance record falls below the guidelines indicated in the corporate employee manual (see excerpt attached), I will recommend that your services at Apex be terminated. I hope we do not reach this point, Sam. Your future could be a bright one here, because your work reflects a good deal of potential. At the same time, this potential will never be realized if you do not bring about the essential improvement required in your lateness and absentee record.

(b) Your work on the Apex project was excellent, despite the fact your absentee record is beginning to exert a very harmful effect on your overall performance record. I sincerely hope that, working together, we can bring about the necessary corrective action. Your supervisor has told me something of the extenuating circumstances involved, but the fact remains that your absenteeism places an extra and unfair burden on your co-workers. We cannot let this situation continue without taking appropriate action. If absolutely necessary, we would consider recommending that you be placed on part-time status. I am not at all certain this would be the best approach, nor am I certain the recommendation would be approved. I would like you to give very serious thought to this problem, and be prepared to discuss your preferred solution with me on Friday of this week at 10 A.M. in my office.

(c)   We appear to be heading toward an action I do not want to take, but will, if it becomes absolutely necessary. Your lateness record (see copy attached) is having a harmful effect on your future here, and on other staff members as well. You have been warned about this, but the fact you have been late seven times so far this month is a clear indication the warning has had little effect. While I can issue this final warning to you, I cannot provide the preferred solution. That must come from you, in the form of immediate and sustained improvement in your attendance record. At the end of next month I would like you to meet with me to discuss one of two things: your continued advancement within this organization or termination of your employment. Your lateness record between now and that time will determine which subject we discuss.

• *INFORMING EMPLOYEE OF PROBATIONARY STATUS*

General Guidelines

When a staff member clearly demonstrates an inability or unwillingness to deliver satisfactory work performance, the time is past for subtle or informal methods. A direct approach is required, one that offers no ambiguities and takes into consideration the legal implications that may be involved. The problem, and the potential consequences, must be spelled out in clear, specific detail. With equal clarity, the employee needs to be told what action he must take to avoid termination of employment. When the message is conveyed initially via a conversation, it must be quickly followed up with a written memo or letter that documents the key factors involved. Memo 6-5 or the "Alternate Approaches" will simplify this action.

Memo 6-5

March 28, 19—

TO: Matthew Bowen

FR: William Brennan

Termination of an employee's services is never a pleasant assignment. I regret the need to send this memo to you, but for all
concerned it's essential that we provide a written record of the
conversation we had this morning. Unless your work performance
shows substantial improvement during the next 30 days, Kenyon
Industries will be forced to terminate your employment. As of
now, you are on a probationary status.

We recognize that personal problems affect all of us periodi-
cally. When this occurs, help should be sought so that such problems do not affect work performance to the point where the
individual is unable to function effectively.

It is obvious to us, and I believe by this time it is obvious to
you, that professional help is needed to help you resolve the
problems that are affecting your performance. The primary reason
we are not taking action immediately is because of the caliber of
your work in the past. This previous record will not be overlooked. At the same time, it is in no one's best interest to let
the present situation continue without a realistic, practical
solution. Your work habits, particularly within the laboratory,
are simply not acceptable. The list of infractions that is attached to this memo details each point covered during our discussion earlier today.

Once again, you have 30 days in which to turn this situation
around. If at any point during this time period it becomes obvious you are not making the appropriate degree of progress in correcting the deficiencies outlined on the attached list, the
termination procedure will be put into effect immediately.

Matthew, we do want to help you. We can't do this if you do not
help yourself. On Friday of this week please join me in my
office at 2:00 P.M. so we can discuss your response to this
memorandum.

Alternate Approaches

(a)   In accord with our conversation earlier today, the attached list provides a written record of the inaccuracies contained in your most recent report to management. It would be difficult to estimate the costly results that would have occurred if we had acted prematurely on your recommendations. It is only because of your previous work record that we did not suggest immediate termination of your employment. As I told you this morning, you are now on probationary status. This will continue until we have clear evidence that you have brought about the necessary improvement, as outlined in the attached performance rating. While I do not want to establish an arbitrary time limit, it is in your best interests as well as ours to avoid a protracted time period. Over the next several weeks I would like to meet with you each Friday to discuss the status of your work.

(b)   One of the most serious actions a manager can take is to place an employee on probation. It is distressing to all concerned. I find it particularly regrettable in your case because of the earlier (and far superior) work you did on the RX100 project. In accord with our discussion of yesterday, the facts require us to place you on probation until June 15. An additional copy of your most recent performance rating is attached. You will notice it has been initialed by the chief engineer. This rating puts into sharp focus the improvements that are necessary in your work. You will receive another performance rating on June 15. I sincerely hope you succeed in bringing about these improvements. I will not send a copy of this memo to the Personnel Department until we know the results of your efforts. If you cannot or will not effect these improvements, I will recommend termination of employment.

(c)   Management has received several strongly worded complaints about the startling increase in product defects during the past two months. As you will note in the attached review of these complaints, the problems have resulted, almost without exception, from errors made by employees under your supervision. This resulted in the review of your work that was conducted last week. The outcome of that review is as disturbing to me as it will be to you. After you review the copy that accompanies this memo, I hope you will understand we have no recourse but to place you on immediate probation. This will continue until one month from today, when I will meet with you to discuss the status of your work. Very sincerely, Tom, I do not want to take more drastic action. Whether I am forced to do so will depend on the progress you make between now and June 30. If you have any questions at all, or would like to discuss this further, please call me on Ext. 2457.

● *TERMINATING STAFF MEMBER'S EMPLOYMENT*

General Guidelines

The final step, referred to in the preceding memo, takes place when it becomes necessary to send Memo 6-6. Viewed positively, the action becomes one that is in the best interests of all concerned. From a career standpoint, few people enjoy standing still or remaining in a situation that offers no hope for individual progress. When this point is reached, the sooner corrective action is taken, the better. Brevity and directness are again the paramount considerations, with compassion being shown in most situations. Recrimination serves no useful purpose, and if you can direct or assist the individual in pursuing another more appropriate career avenue, do so.

Memo 6-6

March 28, 19—

TO: Allan Dyer

FR: Oscar Reed

This will confirm the conversation we had earlier this afternoon. I'm sorry it has become necessary to terminate your services. However, your pay will continue for two full months. This severance payment is based on your period of employment.

The decision to take this action was based on a factual review of your work performance, as reflected by the enclosed Performance Rating. It was not a hasty or ill-conceived decision. You have repeatedly been asked to put more effort and willingness into your work. There was never any doubt that you had the potential to do a competent job, but your last three reports were not only late, when they arrived it became obvious all three contained incorrect information and lacked essential data that should have been included. All this reflected not only on the reports themselves but also on your difficulty in measuring up to the work standards required by the Quality Control Department.

```
Perhaps it would be a good idea for you to consider job possibil-
ities in other areas of engineering. If you'd like to explore
this further, I'd be glad to discuss it with you. Your last day
on the job will be April 12, and I do wish you success in quickly
finding work more suited to your abilities.
```

Alternate Approaches

(a)   Under the circumstances, I can understand why the action we discussed this morning came as no surprise to you. While our earlier conversations over the past few months laid the groundwork for termination of your employment, it nevertheless represents a situation both of us can learn from. You now have the opportunity to devote all of your time to the search for a position that is more closely allied to your interests and abilities. If I can be helpful to you in this effort, please let me know. The severance pay we discussed has been approved, and your last day on the job will be May 12. The Personnel Department, in response to my request, is investigating the possibility of other assignments for you within the organization. When and if something develops, Personnel will be in touch with you. In the meantime, I wish you every success in locating a position that will more fully enable you to achieve your long-term career goals.

(b)   Regrettably, I must inform you that your employment with Apex will terminate on June 1. Although an intensive effort was made to locate another, more suitable, position for you on the engineering staff, this search was not productive. The only position available was on a much lower level, and this is not a time for you to move backward on your career path. On the contrary, this action frees you to concentrate solely on career advancement opportunities. During our discussion this afternoon (when I will give this memo to you), we can review your career objectives and ways in which I might be of some help.

(c)   The recent merger of Acme and Benton Corporation has created a large pool of employee talents that, in many cases, duplicate one another. It is an understatement to say this is most unfortunate, but several employees in our firm will have to be released between now and the end of next month. I wish there were a way I could break this news more gently, but with the deepest regret I must inform you that your position is one that is being eliminated. Anticipating this, we made a concerted effort to locate a transfer opportunity for you. As of today, there is simply nothing appropriate available. We will continue to pay you through the end of August, but your last day on the job will be June 30. You may be sure we will give you an excellent reference, and I would like to meet with you this afternoon to discuss other ways we might be of help.

● *EXPRESSING APPRECIATION FOR ACTION BY*
  OUTSIDE CONTACT

General Guidelines

An effective way to continue, and enhance, a business relationship is to recognize superior service when it is performed. That's the first step. The second, of course, is to acknowledge the exceptional service in the form of a brief note of appreciation. Letter 6-7 reflects the warm, friendly tone to use in situations of this type. By extending your appreciation in writing, you increase the likelihood it will have a longer-lasting effect.

Letter 6-7

ROBINSON'S ELECTRONIC DESIGN, INC.
717 Adolph Street
Monterey, California 00000

May 1, 19—

Mr. William Simon, Manager
Barclay Hotel
Culver Avenue
Richardson, California 00000

Dear Mr. Simon:

For the past year we have been holding our monthly meeting in the Banquet Room of your hotel. Now that we are going into our second year at this location, it seems a proper time to compliment you and your staff for the excellent service and food we have had at each one of our meetings.

Please extend to all your employees who work in the Banquet Room, and to your chef and his staff in the kitchen, our sincere thanks

and appreciation. We look forward to once again enjoying your
fine services in the coming year.

Sincerely,

Alternate Approaches

(a)   Your delivery of the X100 spare parts arrived exactly when you said
they would. I won't bore you with the details on how much time and effort
this saved us in making the necessary repairs, but I at least want to thank
you for being so dependable. It's a refreshing change from some of the
experiences we've had with other suppliers. I'm sending a copy of this letter
to our purchasing director and look forward to doing business with you in
the future. Thanks again for being so cooperative.

(b)   Most people are gratified when an authority in a given field agrees
with an opinion they've expressed, and I am no exception. Your support of
the position I maintained during our last management conference was
sincerely appreciated. Subsequent events have convinced me it was your
agreement with my recommendation that helped turn the tide. Because of
your affiliation with an outside agency, you were able to offer a more
objective view at a time when the observation was most needed. Thank you
very much, Dr. Baker. As we proceed on the project involved, I'll be glad to
keep you posted on developments.

(c)   Since this was the first experience I have had with an outside research facility, I was not prepared for the extraordinary data you provided.
To say that I was impressed would be an understatement. On the basis of
the well-organized and documented statistics you provided, we have decided not to go ahead with the Harrington project. This became a relatively
easy decision to make, after we had gone over your report in detail. Our
staff appreciates the really fine effort you put into this, Dr. Mergner. Rest
assured we intend to take advantage of your excellent services in the future.

● *COMPLIMENTING PRESENTATION BY*
   SEMINAR PARTICIPANT

General Guidelines

Appreciation of the type expressed in Memo 6-8 can go a long way
toward making professional relationships more pleasant—and often, more

productive. It is usually best to express your thanks as soon as possible after the occasion presents itself. A brief note is little enough to give in return for the extra effort that some conscientous or generous associates extend. To make it more factual, mention a few definite things that pleased you. When appropriate, also make it clear that the appreciation is extended on behalf of the people you represent.

Memo 6-8

June 15, 19—

TO: Gordon Newell

FR: Cary Spitzer

On behalf of everyone who attended the "Advanced Aircraft Systems" session at the recent AIAA Aerospace Engineering Conference, and on behalf of the conference management, I want to thank you for your presentation on "Reliability, Maintainability, and Cost Evaluations of Digital/Electric Aircraft Controls."

Your presentation was properly focused, expertly delivered and created a very favorable image for your organization. You maintained your usual high standards and were able to offer previously unpublished data and analyses. It was one of the best attended sessions at the conference, and the nature of the questions reflected strong audience interest.

Again, thank you for your participation, Gordon. I look forward to our continued association as we pursue our common goal of safer, more efficient aircraft.

Alternate Approaches

(a)   Your remarks at the March 12 Engineering Association meeting added up to a masterful presentation. I'm sure you would be as gratified as I was to hear the many favorable comments expressed by the membership. Some of the research data you shared with us on new radiology developments both surprised and impressed us. All in all, it was a most informative and interesting talk, and one that offered practical benefit to all those who attended. Thank you very much, Frank. Whenever you are in this area again, I hope you'll give me a call. The least I can do is take you to lunch, and I'll be glad to do so whenever it's convenient for you.

(b)   Having talked with our program chairman regarding the regional meeting on May 4, I expect you'll soon be getting a note from him on your participation at this meeting. Frankly, it was the highlight of the entire evening. I have seldom seen our membership so attentive, and so interested in the remarks by a guest speaker. I am simply expressing the consensus of opinion when I tell you it was a superb presentation. It might be said that every silver lining has a cloud, however, and in this case it undoubtedly means we'll be calling on you for another presentation in the future. A performance of that caliber deserves an encore, and we'll be ready whenever you are. Once again, Bob, sincere thanks for the obvious thought and effort that went into your remarks.

(c)   The entire engineering staff at Arlington Mfg. joins me in expressing our appreciation for your splendid presentation during our annual conference on March 24. Your remarks were well documented, well organized and were delivered in a way that captured, and held, the attention of all those who attended. As a matter of fact, we would like your permission to include a verbatim transcript in our corporatewide newsletter. This goes beyond the confines of our corporation, and I've enclosed an earlier issue for your review. Provided you have no objection, we'll plan to include this in the June issue. If there are portions you'd like to omit, just send me an edited version, and we will go with that. Thank you very much, Dr. Remington.

● *RECOMMENDING DISCIPLINARY ACTION*

General Guidelines

Complaints of the type referred to Memo 6-9 are difficult enough when one of your employees is involved, but the situation is compounded when it involves someone in another department. It is important to present a factual review of the problem. Your position must not only be clearly stated, but wherever possible, you should spell out the damaging effects that will result if the problem is not corrected. If earlier attempts have been made to counsel the employee involved, this should also be referred to.

Memo 6-9

February 24, 19—

TO: Helen Jennings, Cafeteria Supervisor

FR: Carol Overstreet, Engineering Staff

RE: Wilma Vernon, Cafeteria Worker

This memo requests that you take an appropriate form of disciplinary action with respect to Wilma Vernon.

Wilma has been employed at our branch cafeteria for three years and has been a continual source of aggravation and discontent, not only among her co-workers but among corporate staff employees as well.

For the past six months she has been the central figure in virtually every unpleasant and disruptive situation involving cafeteria workers and corporate staff. She seems unable to get along with any other worker—so it is not merely a case of bad feelings between her and only one or two others.

The cafeteria is an area critical to the operation of our offices because of the distance involved in traveling to other locations offering food service. While I am wholly in favor of helping employees grow and improve in their positions, I think we must identify the time when such counseling has failed and think about other alternatives. It is not fair to other workers who are trying to run a quality service to be hampered by a single malcontent.

I hope you agree that termination or some specific form of disciplinary action is appropriate. Please let me have your comments.

Alternate Approaches

(a)   Tom, I had hoped to avoid writing this memo, but an incident in the Production Department this morning forces me to bring the problem to your attention. I believe William Benton on the maintenance staff reports to you. Over the past three months we have documented 5 occasions when problems developed on the assembly line, and no one, but no one, could locate William Benton. In each instance he was unable (or unwilling) to explain his absence satisfactorily. The delays in production that resulted from these incidents range from 15 minutes to 1 hour and 20 minutes. This is costing us time and a substantial amount of money. Some form of disciplinary action is clearly needed. Would you please let me know what you plan to do about this?

(b)   We have talked before about Andy Warren's carelessness with lab equipment and his generally poor attitude with others on our staff. The counseling sessions I have had with Andy seem to bring an improvement that lasts for a few days; then he reverts back to his usual lackadaisical manner. Per your request, I will not take stronger action on this without discussing it first with you. But I am now recommending that we consider assigning him to a lower-level job. If he refuses to accept the assignment, then termination appears to be the only remaining answer. I have documented the problems we've had with him (see attached) and would like to discuss this with you as soon as possible.

(c)   You have asked for my recommendation regarding problems in the Quality Control Department. Regrettably, I have to say that the key factor here, and the major problem we should resolve quickly, lies in the supervisor himself. For whatever the reason, and I think I know what it is, John Emory has lost control of this department. He has neither the skill, nor the intuitive ability to supervise people without trying to intimidate them. As a result, morale in this department is at a low point, and it isn't about to improve until a new and more people-oriented supervisor is brought in. My opinion is that John would probably be relieved, in more ways than one. He's uncomfortable with the increased responsibility he was given, and it's quite likely he would welcome reassignment to a less demanding position. I have attached a list of transfer possibilities and would like to discuss this with you. Please let me know when it's convenient.

● EXTENDING APPRECIATION TO A VALUED CLIENT

General Guidelines

Sending an occasional note like the one illustrated in Letter 6-10 is a thoughtful gesture and one that could bear dividends in the future. While it is more a good customer relations letter, and not strictly sales oriented, the obvious objective is to ensure a continuing relationship with an important client. No need to embellish the facts; simply express your appreciation and the hope it will continue to be a mutually satisfactory relationship in the future.

Letter 6-10

ARLINGTON ENGINEERING SERVICES, INC.
1211 Ames Boulevard
Macon, Virginia 00000

November 26, 19—

Mr. Thomas Brown
Apex Laboratory Equipment, Inc.
412 Keystone Avenue
Hampton, Virginia 00000

Dear Mr. Brown:

Several years ago your organization first favored us with a request for our services. Four years ago, to be exact. We wish to mark this important anniversary by telling you how much this relationship has meant to us.

Our organization, at the time you first contacted us, was then a small group—but with great ambition and a determination to grow. Growth has indeed occurred since then, and it has been due in large measure to the confidence and loyalty of the many friends we are privileged to serve.

I am happy to include you among these friends, and hope that as the years go by, we can continue to make our services even broader and more useful to your organization.

Our sincere thanks are extended to you, Mr. Brown, along with every good wish for a most enjoyable holiday season and a healthy, happy and productive New Year.

Cordially,

Alternate Approaches

(a)   It is with a great deal of pleasure that I send this letter, marking the third anniversary of our association. The fact you have continued to use our services over the years means a great deal to me, not only in a business sense, but also because of your courtesy and understanding on so many occasions. This brief note is simply to express our appreciation. You may be certain we'll do all we can to see that this cordial relationship continues in the future.

(b)   We sincerely hope that our designs have been satisfactory to you over the past year. Rest assured we will continue to make every effort to maintain and improve our services so that our pleasant and productive association will continue. Time seems to move so rapidly I didn't want our "first anniversary" to go by without thanking you for your patronage. It is a pleasure to be of service to you, Mr. Rand.

(c)   Client relationships, of course, determine both the present and future success of any professional organization. That is why we make a constant effort to maintain not only a productive relationship with those who use our services, but we want that relationship to be cordial as well. Our chief engineer recently told me that it has been five years since you became one of our clients. Not only has the association lasted five years, but Bob Egan also pointed out that your organization has consistently been reasonable and cooperative during the entire period. This is indeed gratifying, and I just want you to know we appreciate the opportunity to be of service to you. Thank you very much, Mr. Grant.

● *COMMENDING STAFF MEMBER FOR EXCELLENT PERFORMANCE*

General Guidelines

Sending a note of commendation to a deserving employee is a simple but effective way to show that you recognize and appreciate his or her performance on the job. Memo 6-11 deals with recognition of a particularly useful idea submitted by the staff member, but other types of contributions could also serve as the basis for such a memo. You will want to indicate the specific reason for the memo and make certain a copy is sent to the appropriate supervisor. A copy should also be made part of the individual's employment file.

Memo 6-11

March 28, 19—

TO: Paul Wainwright

FR: Richard Thornton

Paul,

I'm pleased to inform you that the Management Committee has authorized me to write this letter of commendation for the adoption of your proposal to fabricate inhouse Part No. 16ZP-738, Details 7 and 34, instead of buying these items from vendors.

Implementation of your suggestion effected direct savings to the company because of the difference between outside purchase costs and inhouse manufacturing costs.

The inventiveness and initiative you've demonstrated are greatly appreciated. In addition to the award that accompanies this memo, you can be certain management will give special consideration to ideas you submit in the future.

I'm sending a copy of this memo to the vice-president of engineering, along with copies to the chairman of the Management Proposal Committee and the personnel director.

Richard

Alternate Approaches

(a) At a recent management conference, your supervisor told me of your contributions to the X100 design. I was both surprised and gratified to hear this—surprised because you were assigned to this project only recently, and gratified because your ideas will help us complete this project on schedule. It's great to know you have so quickly become a key player on the design staff team. You have our sincere appreciation, and I hope you will continue this exemplary work performance. Both your supervisor and myself will see that your design improvements on the X100 are made part of your personnel file.

(b) Because of your excellent quarterly report, we have decided to expedite a plan that was developed shortly after your last performance rating. Specifically, your assignment to the research staff has been approved, and we'd like you to assume this position on March 1. It's always a pleasure to convey news of this type, but especially so in your case. Your progress toward this new assignment has been both steady and impressive. This action is a tangible expression of our appreciation, and all of us recognize you have earned it by your hard work and a superior job performance. The formal announcement will be made at the next staff meeting, and your new responsibilities are spelled out in the attached job description. Please meet with me on Thursday morning at 10 A.M. so we can discuss this in more detail.

(c) Congratulations. I'm delighted to tell you your plans for the Manchester development have been approved. Frankly, I had anticipated a struggle in getting the committee to select your proposal, but, fortunately, they were quick to recognize the superior caliber of your design. This sets off a chain reaction in the sense that bids can now be secured earlier, and we should be able to break ground within the next two months—all because of the thorough, well-organized proposal you submitted. This achievement will be made part of your employee record, and please accept my thanks for a job very well done.

● *RECOMMENDING COLLEAGUE FOR EMPLOYMENT OR REASSIGNMENT*

General Guidelines

Most letters of recommendation deal with employees going to another firm, as illustrated by Letter 6-12. However, an equivalent endorsement may also be put in memo form when recommending a reassignment for one of your staff members to another part of the organization. Make your statements specific; avoid generalities. Key factors to cover include the person's full name, the duration of your association, your relationship to the individual, and several concisely stated reasons why you feel the person has superior work habits and skills.

Letter 6-12

WILLIAMSPORT MANUFACTURING CORPORATION
2800 West Fourth Street
Williamsport, Pennsylvania 00000

January 19, 19—

Dr. Merrill Baker
Manager, Engineering Staff
McLoughlin Corporation
Markham, Oregon 00000

Dear Dr. Baker:

Leona Sayre has been a design engineer on our staff for the past four years. During that time, I have had an opportunity to observe and evaluate her work as well as her potential for growth in her field.

She has excellent rapport with her co-workers and shows a genuine interest in them as individuals, while constantly searching for ways to make her relationships mutually productive in a professional sense. My observation was that her colleagues liked her, respected her, and were pleased to have her on the staff.

Leona is a frank and outgoing person, possessing great technical skill, along with a cooperative and friendly personality. Her appearance is beyond reproach, and her attitude consistently positive and optimistic. I've found her very receptive when assigned extra work, and she frequently volunteered for special assignments when the need arose.

Mrs. Sayre is leaving our organization because her husband has been transferred to your state. If there had been any way she could be retained, I would most assuredly want to keep her on my staff. She is a capable and dedicated professional engineer and an asset to our organization. It is difficult to find a design engineer with her talent, ambition, and drive, and I recommend her without reservation for the position she is seeking.

Sincerely,

Alternate Approaches

(a)   Thomas Manning has progressed at an accelerated pace during the past few years, and justifiably. He has consistently displayed a "can do" attitude, enjoys the friendship and esteem of his peers, and has contributed many of the smoothest running systems in our division. You can probably understand why I have mixed feelings about this memo of recommendation. I hate to lose Tom. It is only because he has reached the top level in his present assignment. Regrettably, we have no appropriate vacancy on a higher level, and nothing on the horizon that would match his skills. His personnel file is attached, and I am more than willing to supply additional data if you need it.

(b)   I consider it a pleasure to respond to your letter of inquiry regarding Alice Carlin. Mrs. Carlin served as one of the key people on our research staff for over four years. Even though she relocated to your area several months ago, we still miss her. Her reports, many of them on extremely complicated subjects, were always submitted on schedule and were written with great clarity and attention to detail. I cannot recall even one occasion when we were not satisfied (I should say delighted) with her work. It is not only my opinion that I express. Her co-workers, without exception, liked and respected Mrs. Carlin for her professional skill and for her pleasant manner in dealing with people. You will indeed be fortunate if she becomes a member of your organization. I have no reservations at all in strongly recommending her to you.

(c)   This letter of endorsement is in response to your reference request concerning Harold Gray. Mr. Gray is an excellent and highly skilled draftsman who successfully completed many assignments during his three years on our staff. As a matter of fact, any assignments that were particularly difficult were always given to Mr. Gray. Not only were his technical skills far above average, he managed to go the extra mile on many occasions by offering ideas that went beyond his responsibilities. He won six suggestion awards while with us, for ideas that helped effect substantial savings in operating costs. He continued his studies while employed here and, upon securing his degree, decided to pursue a career in digital avionics. That is what led to our separation, but if he ever wanted to return, we would be delighted to have him.

A WORD ON STYLE

Pay Attention to Transitions

English teachers call it coherence. It is the "sticking together" of sentences and paragraphs, and it may be the hardest thing for anyone—even the most experienced writers—to do well.

When you start to write a letter or memo, you already know your subject thoroughly. Your reader does not know it half so well. Therefore, you are going to have to join together all the links in the chain, to fill in the gaps, and you are likely to find it difficult.

Most audiences for your letters, as for other kinds of writing, are a mixed group. Some know a great deal about your subject, some know nothing. An informed reader can supply an astonishing number of connections, even in letters that make very few connections on their own. An uninformed reader, of course, is lost.

English teachers will tell you to pay attention to the howevers, the therefores, the thens, and the nexts. Their advice is good—as far as it goes. The deeper problem is one of clear and sequential thought and logic. The transitional words, important as they are, require a firm grounding in reality. Much the tougher (and much the better) transitions are invisible. They are those that carry the reader from sentence to sentence, from paragraph to paragraph, without the intrusive use of signals.

Chapter 7

PERSONAL LETTERS AND MEMOS TO PROFESSIONAL CONTACTS

There is often a fine line between a straightforward business letter and the "personal" letter or memo. There are also occasions when the circumstances justify blending the two. This might occur when, toward the end of a letter dealing with business matters, you wish to extend a "personal note of thanks" for the reader's cooperation and understanding. While business letters dealing with technical matters are, in a sense, obligatory, the personal letter is usually written on your own initiative and expresses your own personal sentiment. Generally, it is the letter with a friendly, reasonably informal tone that will be most appreciated. On the other hand, an approach that is too effusive could offend the reader. To find the right degree of informality, ask yourself if the person you are writing to is a close friend or an acquaintance you really don't know very well. In any case, your remarks should be focused specifically on the occasion that prompted the letter and not deal with peripheral or extraneous matters. It should be made clear at the outset why you are writing, and the note should not be any longer than it needs to be. This chapter offers a wide variety of approaches you can use when writing personal letters. Most are short, but each one carries the warm, friendly tone referred to earlier.

● *OFFERING CAREER GUIDANCE TO A STAFF MEMBER*

General Guidelines

Receptivity to advice depends largely on the spirit in which it is given. An employee who demonstrates potential in a given field is usually the type who will welcome, and appreciate, career guidance. Particularly when it stems from a genuine desire to accelerate the employee's progress toward a specific goal. Memo 7-1 is based on the supervisor's overall view of three factors: the particular skills of the staff member, objectives facing the department, and avenues of opportunity for the employee in question. The more concisely the advice is stated, the more likely it will be considered seriously—and acted on.

Memo 7-1

April 22, 19—

TO: Frank Mitchell

FR: John Benchley

Frank,

It was quite a year, struggling to overcome the many difficulties all of us had to face. I want you to know that I feel you deserve much of the credit for our turnaround in production capability. As a matter of fact, your performance was such that I'd like to offer a bit of guidance regarding your future on our staff.

It is my intention to consider you as a possible replacement for Willard Benton when he is reassigned to the headquarters office. I would be reluctant to do this, however, unless you developed greater familiarity with systems and procedures in effect in that department. First, of course, we must determine if you are interested in the assignment. I suggest you meet with Willard for an in-depth picture of the responsibilities involved. Then, I would like to meet with both you and Willard to discuss this further.

```
In the meantime, I want you to know we appreciate your fine
performance in the Production Department during the last several
months.

John
```

Alternate Approaches

(a)   I'm glad we had a chance to spend some time together during the performance evaluation session last week. You gave me a clearer picture of the assignments you'd like to work on in the future, and I'd like to help you achieve these objectives. First, there's no question you have the necessary technical skills. Both your formal engineering background and your performance on the job convinces me of this. Since your goals will involve leadership responsibilities, the primary suggestion I have deals with this aspect. Specifically, you might consider a course or two on management, particularly as it relates to the supervision of people. I have attached some descriptive material on courses offered by the local university and would be glad to discuss this further with you if you are interested.

(b)   As you know, we've been giving a good deal of thought lately to the ways in which we might explore the practical use of robotics in the Apex plant. A number of our manufacturing processes could benefit from this exploration, and the search could also play a part in your future. I was particularly interested in your remarks on this subject during the last supervisor's meeting and have an idea I'd like to share with you. The Matson Corporation has done considerable research in this area and recently completed the installation of several devices in its Worthington plant. I know Matson's production manager quite well, and I believe he could save us a lot of time if we approached him on this. His name is Bob Kern. I'd like to call Bob and ask him if you could spend a couple of days talking to him and his staff about this. Give the basic idea some thought, and then call me when you're ready to discuss it further.

(c)   You probably know by now that I have a genuine interest in your future, and particularly in your advancement to a higher level on our staff. This prompts me to make a suggestion I hope you will consider seriously. Bob Randall informed me yesterday that he plans to retire at the end of this year. You will be one of the people considered for that position. The problem, as I see it, has to do with your lack of experience in computer technology. I have asked Bob to explain to you the importance of a working knowledge in this area, and he has agreed to do so. Once you recognize how important it is, I hope you will consider taking a crash course, such as the one Bob mentioned to me when I discussed this with him. First, meet with Bob (the sooner the better). Then, let's get together so we can discuss this opportunity in more detail.

● *EXTENDING CONGRATULATIONS TO A COLLEAGUE*

General Guidelines

One of the more pleasant writing assignments involves congratulating someone on an important achievement. Use expressions that would be natural in a conversation with the reader. As illustrated by Letter 7-2, don't be excessive in your praise or smother the occasion with overblown phrases. The sincerity of your remarks should come through clearly, and the tone should be geared to the personality of the recipient. It is seldom necessary or advisable to make such a note longer than two or three paragraphs. State the reason for your congratulations in the first or second sentence, followed by comments that link the person and the occasion. A memo or letter of this type can do a great deal to strengthen and continue a worthwhile professional relationship.

Letter 7-2

WORCESTER MANUFACTURING COMPANY
222 Ellen Drive
Chicago, Illinois 00000

April 22, 19—

Mr. Michael Allen
412 Kent Street
Arco, Kansas 00000

Dear Mike:

Although we have long been separated by job changes and geography, I just want to send congratulations on your new position as

chief aeronautical design engineer. You must be delighted to realize you've been so successful in the field that has held your
interest for so long. Around here, we still miss your cooperative
personality and your innovative solutions to many of the problems
we faced.

Please stop by any time you're in town, and once again, our
warmest good wishes on your new assignment.

Cordially,

David Bradshaw

Alternate Approaches

(a)   The good news of your appointment to product manager was passed
along to me by our mutual friend, Bob Chandler. I could not have been
more pleased, Tom. I have a pretty good idea of how long and hard you
worked for this promotion, and am delighted the effort paid off. You have
some busy days ahead, but I'm sure you wouldn't have it any other way.
Here's hoping the job will be challenging enough to be interesting, but not
heavy enough to be a burden. Either way, I'm sure you'll be able to cope
with it. Our very best wishes for your continued success.

(b)   What a nice surprise it was to read the announcement about you in
the corporate bulletin of May 12. I had no idea you were being considered
for the Prescott Award, but I can tell you I was delighted to hear you
received it during the last annual meeting. Ordinarily, I would have been
there, but a trip to our branch office in Boston prevented me from attending. Nevertheless, it was great to hear the good news, and I hasten to add my
congratulations to the many others I'm sure you've received. Now, I'm
looking forward even more to my next trip to the West Coast. You can be
sure I'll stop by your office so I can take you out to lunch. It will be great to
see you again.

(c)   My hearty congratulations on your election to the presidency of the
Western Engineering Association. Your election is an earned tribute from
your colleagues and also provides recognition of the outstanding work
you've done for the Association and for your profession. I know that with
you at the helm the Association will continue to grow and prosper. Along
with the other duties involved, it seems likely you'll want to make an
occasional trip to various chapters within the Association. Please accept

this as an invitation to make ours one of the first regional groups you visit. This will enable me to extend congratulations personally and renew a valued friendship. Best wishes to you, Frank. I look forward to our next meeting.

● *LETTER OF CONDOLENCE TO WIDOW OF EMPLOYEE*

General Guidelines

Of all the times when sensitivity and compassion are called for, a death in the family has to be considered one of the most important. Condolence letters should be straightforward and reflect a sincere concern on the part of the writer. A gentle touch is required. Your thoughts will be less difficult to phrase if they are written from the heart. Letter 7-3 provides one example, and others in this chapter can be adapted to fit different situations. Note that each one is relatively brief, while containing variations based on the writer's relationship with the bereaved.

Letter 7-3

CANTON MANUFACTURING, INC.
722 4th Avenue
Detroit, Michigan 00000

April 4, 19—

Mrs. Alice Robertson
416 Keystone Drive
Alston, Michigan 00000

My Dear Mrs. Robertson:

Everyone on our engineering staff was profoundly shocked and saddened by the sudden passing of Bob.

Although sympathy can only be a small consolation, even from the hearts of us who share your sense of loss, I want you to know how

much we'll miss his presence here. He was respected and admired
by everyone who worked with him.

While we cannot lessen your sorrow, you may be sure that each
individual of our company joins in this expression of our deepest sympathy.

Very sincerely,

_____

Alternate Approaches

(a) You may be certain that everyone on the engineering staff shares in
your grief at this very sad time. All of us had the highest regard for Tom,
and we shall miss him too. In addition to his many fine qualities, he had a
remarkable ability to cope with difficult situations, and we saw so much
evidence of that during the time we were privileged to have him on our
staff. Those of us who knew him so well will try to emulate that quality
now. Please know that our thoughts are with you and that we stand ready to
help in any way we can.

(b) Although we have never met, I feel that my friendship with Bill was
so cordial that it extended to his family as well. Please accept my most
sincere condolences on the loss of your husband, and my friend. Although
the nature of our work brings us in touch with many people, there is no one
for whom I had a higher regard than Bill Watson. Our association went
beyond the limitations that usually apply to business relationships. His
cheerful manner and outgoing personality earned him a great many
friends, and I feel privileged to have been one of them. I realize your fine
family will help to console you during this time of great sadness, but I stand
ready to help too. If ever I can be of service, Mrs. Watson, I hope you will
let me know.

(c) All of us were shocked and deeply saddened by the news that
reached us this morning. Dr. Johnson was a good and gracious friend to
members of our staff, and we share in the great sense of loss you must be
feeling. Please know that our thoughts and prayers are with you. Having
worked with Dr. Johnson for several years I can tell you he was not only
highly regarded here, he was an inspiration. Although he left a great legacy
of good work, we shall miss him deeply. Perhaps the many good memories
will help sustain us at this very sad time, and we just want you to know that
we share in your sorrow.

● *EXPRESSING GOOD WISHES TO ILL STAFF MEMBER*

General Guidelines

   In a letter expressing concern over an illness or similar misfortune, tact and sincerity are the most important qualities. As in the case of Letter 7-4, a solicitous letter of this type should also carry a feeling of warmth to the recipient. Other guidelines would include the need to keep the letter brief and the tone encouraging. Don't philosophize, or overly dramatize your reaction to the illness or misfortune. If the relationship and circumstances are such that you wish to offer help, the usual place for this would be in the last paragraph.

Letter 7-4

TRANSATLANTIC AIR CARGO
3900 Juniper Avenue
Kingston, New York 00000

February 17, 19—

Mr. Peter Marshall
850 Baylor Avenue
Bay View, New Jersey 00000

Dear Pete:

Since we were together just a few days ago at the committee meeting, I was deeply shocked when I learned this morning that you are in the hospital. However, it was a relief to hear that the operation was successful and that you are on your way to a rapid recovery.

While I know the hospital is not the most pleasant place to be, I hope you will try to rest and take it easy for as long as necessary before returning to work.

Please know that our thoughts and best wishes are with you, Peter.

If anything develops that I might be helpful with, just let me
know.

Very best wishes,

Ronald Carter
Chief Engineer

Alternate Approaches

(a) I called your office today to set up a luncheon appointment and was astonished to hear you are in the hospital. I was very sorry to hear this, Jim. Perhaps you're not certain at this point how long you will be there, but I'll keep in touch with your office and try to get over there to see you whenever you're ready to have visitors. We miss you already. Please let your wife know we'd be glad to help in any way we can. In the meantime, don't worry about the job. You've got a lot of friends here who will help out, and all of us send our warmest good wishes for your rapid and complete recovery.

(b) We had the monthly staff meeting yesterday and Bill Barnett told me of your illness. I was shocked, of course, but knowing you, I'm certain you'll somehow manage to lick this thing in the shortest time possible. If it had to happen, it is good that you had already trained your assistant to take over in emergencies of this type. You've helped to raise Tom's level of confidence and skill to such an extent there is no need for you to worry about things going undone. The more important things that only you can do will simply have to wait until you return. In the meantime, the most important thing right now is for you to get better. Please know that our thoughts are with you and that you have all our best wishes for a speedy recovery.

(c) Your wife called this morning to tell me that you are in the hospital. It's an understatement to say I was sorry to hear this, and you can bet all of us are hoping you will only be there for a short time. Just one short note about business: you'll be glad to know your recommendations on the Watchhaven project were all approved. I know how much thought and effort you put into this and pass along the good news only because it might contribute something to your recovery. Now, of course, we will simply have to wait for you to put the plans into effect. We would not, however, want you to rush things in order to do this. Your health is, by far, the most important consideration right now. We're thinking of you, Tom. June will call me when the doctors say I can stop by for a visit. I hope that will be soon.

- *CONVEYING SYMPATHY TO FAMILY OF*
  DECEASED COLLEAGUE

General Guidelines

A letter of sympathy to the family of a deceased staff member is a gesture of respect for the colleague, and for the family as well. As with the earlier letter to an employee's widow, Letter 7-5 conveys a feeling of empathy and expresses consideration and concern at a time when it is most needed. While respect and sincerity are prime prerequisites, so is brevity. Phrases such as "words cannot express" or "there is nothing anyone can say" should be avoided—mainly because they seem to express a sense of helplessness at a time when the reader needs help. When necessary, state your former relationship with the person being mourned. The circumstances will determine if you should extend an offer to help.

Letter 7-5

HARVIE MANUFACTURING CORPORATION
812 Syracuse Boulevard
Darby, New York 00000

March 31, 19—

Mrs. Ronald Emerson
22 Keynes Lane
Greenville, New York 00000

Dear Mrs. Emerson:

Everyone on the engineering staff joins me in expressing our deepest sympathy to you and your family.

We came to know Dr. Emerson well these last few years, during his visits to our firm. His wisdom and counsel meant a great deal to me. He was a personal as well as a professional friend, and we will indeed miss him very much.

`Please know that our thoughts are with you and that we stand`
`ready to help in anyway we can.`

`Sincerely,`

Alternate Approaches

(a) It was with heartfelt sadness that I learned of the news regarding your son. Bill was part of our engineering staff for only a relatively short time, but it was long enough for us to discover he was an exceptionally fine young man. We worked together on several assignments, and I came to know him not only as a skilled colleague, but also as a warm and understanding friend. Your loss is our loss too, Mrs. Stanton. Please accept our most sincere and deepest sympathy.

(b) We were stunned yesterday to learn of Fred's untimely passing. It would have been twenty years ago next month that Fred became part of our organization. During all those years I came to know him as a warm, understanding friend, and one who will occupy a very special place in my memory. Please know that our thoughts and prayers are with you and that we stand ready to be of help in any way we can.

(c) My path did not cross with Helen's nearly as often as I would have liked. We never worked for the same firm together and met only because we were members of the same professional association. Yet I looked forward to each meeting. I not only liked Helen, I admired her. Our paths crossed more frequently when we worked together on the same committee, and I shall always remember her gentle, considerate ways. Without question, she was one of the most highly regarded members of our association. We shall miss her too, Mr. Baxter, and extend our deepest sympathy to you and members of your family.

● *LETTER OF CONDOLENCE TO EMPLOYEE ON DEATH OF FAMILY MEMBER*

General Guidelines

Letters of sympathy can become even more difficult when offering consolation to someone regarding the loss of a loved one you may never have met. Moreover, laboring over such a letter can make it sound forced and insincere. Letter 7-6 is one that can be adapted to situations of this type. A simple, heartfelt message is the kind that will be most appreciated. Keep the use of "I" to a minimum, focusing your remarks on the reader. As with other letters in this category, brevity and simplicity are paramount factors. Don't heap sentiment on sentiment. Doing so will make the letter sound stilted, if not artificial.

Letter 7-6

WALTER J. REED CORPORATION
860 Bostwick Avenue
Peoria, Illinois 00000

March 31, 19—

Mr. Bradley Wilson
444 Wilkinson Boulevard
Chicago, Illinois 00000

Dear Mr. Wilson:

Please accept our most sincere sympathy on the loss of your son, Jerry. He was much more than a professional colleague to us. He was also a widely respected and admired friend. We do, indeed, share in your sorrow. If there is any way at all we might be of assistance to you, Mr. Wilson, I do hope you'll let us know.

Sincerely,

Alternate Approaches

(a)　All of your many friends here were deeply saddened to hear the news that reached us this morning. While we did not know your husband, we've known you over the years and that alone tells us he must have been a very fine person. Please remember, Helen, you are not alone in your grief. We want you to know that our thoughts, and our prayers, are with you.

Particularly at times like this, our relationship goes far beyond the fact we are professional colleagues. We are friends too. And your friends want to help if there is any way at all we can be of assistance to you.

(b)   You have our profound sympathy on the loss of your daughter, Tom. As I'm sure you know, mere words can be so inadequate at a time like this. Yet, I hope it will be of some comfort to you and your wife to know that all of us here do, indeed, share in your grief. We did not have the privilege of knowing Donna, but that in no way reduces the sadness we feel. Is there something, anything, we can do to help? You have only to ask.

(c)   Moments ago we were informed of your son's fatal accident, and, like all of your many friends here, I find it difficult to express the shock and sadness felt by every member of the staff. Situations like this are so painful, and the emotional intensity so great, that one's true feelings become almost inexpressible. While Bobby was a member of your family, it might be said that we think of you as a member of ours. And we want you to know that you have our most sincere and heartfelt sympathy. You will be hearing from other members of the staff, but there is one thought that could be expressed on behalf of all of us: if there are ways we can be of assistance to you and your wife during this sad time, we do hope you will let us know.

- ## *CONGRATULATING STAFF MEMBER ON A PARTICULAR ACHIEVEMENT*

General Guidelines

As contrasted with Memo 6-11, the congratulatory note indicated in Memo 7-7 offers a more personal recognition of an employee's accomplishment, whether or not it is job related. Many occasions are appropriate for a letter of this type. In addition to educational achievement, there are such things as civic contributions, advancement within a professional association, or publication of an important paper. The tone should be geared to the relationship you have with the individual and to his or her personality. To a rather staid person, straightforward and conservative statements would be in order, while to a long-time and close friend, a warm and friendly approach would be more appropriate.

Memo 7-7

April 22, 19—

TO: Howard Smith

FR: Charles Baker

The long years of study, while holding down a full-time job has
finally led to your well-deserved Masters Degree. My hearty congratulations on an honor I'm sure that did not come easily. This
degree will certainly enhance your opportunities in engineering
and should also increase your value to our staff. Due recognition of this will be given in light of your continued fine performance.

Keep up the good work, Howard. We're proud of you.

Charles

Alternate Approaches

(a) I realize you're not the type of person who would want special
recognition for fulfilling an obligation you assumed, but I'm going to send
this note anyway. The success you achieved as a leader in this year's
United Fund drive was little short of remarkable. It has brought credit to
you and, indirectly, to the organization you are part of. Typically, we
learned of your success through Tom Watkins and not from you. Tom
read the article in your hometown newspaper and sent a copy to me. We're
proud of you, Bill. Working on behalf of others who are so much less
fortunate than ourselves is a noble endeavor. Congratulations on another
job well done.

(b) Although I don't get around to the branch offices as much as I'd like
to, I do, as you know, keep in touch. This memo is a change of pace in that
it deals only peripherally with business matters. Leadership skills are important to any organization, and I was delighted to learn of your recent
election as president of the Newport Rotary Club. To my mind, this
achievement is just as important to us as it is to the membership of your

Club. It's another reflection of your desire to serve—and to lead. Congratulations, Bill. I'm sure the membership will be as pleased with your performance as their new president, as we are with your contributions to our Engineering Department.

(c)   I just heard the good news, Dr. Wainwright, and could not have been more pleased. Of course, we have long been aware of your contributions to laser optics, but to have this recognition reinforced by the national association, well, that just confirms the opinion we've had all along. Your colleagues here at Apex join me in expressing our admiration for your work in this field. Having your paper selected as one of the "10 Best" out of 800 entries from across the country is an honor indeed. Several of us, including those you have worked most closely with here, will attend the awards ceremony next month. We hope to see you there, and look forward to expressing our congratulations personally.

- ## *PERSONAL NOTE OF THANKS TO PROGRAM CHAIRPERSON*

General Guidelines

Attending conferences, seminars, and/or symposiums are often essential activities for anyone involved in a technical profession. All of us have experienced occasions when some programs have been better organized and more productive than others. On those occasions, letting the program chairperson know you were pleased is not only an act of simple courtesy, it has its practical aspects too. For one thing, it will help pinpoint the exact reasons why you felt the session was productive. And this will, in turn, help the organizers plan future sessions with these factors in mind. As with most letters or memos of this type, and as suggested by Letter 7-8, the sincerity of your remarks will be emphasized by brevity.

Letter 7-8

NEW ENGLAND MANUFACTURING ASSOCIATION
910 Albany Avenue
Queens, New York 00000

June 1, 19—

Mr. Robert Thornton
Seminar Productions, Inc.
1210 Metropolitan Avenue
Atkins, Illinois 00000

Dear Robert:

Last week's DELPHI convention in Chicago was most enjoyable,
primarily because of your efforts as program chairman. We not
only had interesting and knowledgeable speakers, but every event
went smoothly and according to schedule.

On a more personal note, my wife wants you to know how much she,
too, enjoyed the convention. She thought the side trips and entertainment you arranged for the ladies were simply outstanding.

It was good to see you again, and I hope we can get together for
lunch during your next trip to this area. I believe we may have a
couple of engineers who could make a worthwhile contribution to
one or two of the seminars you are planning for the future. We
can discuss this in more detail during your next visit.

With best regards,

Alternate Approaches

(a)   I almost missed last week's regional conference because of a planned
business trip. Fortunately, I decided to postpone the trip because I was
curious as to how you were going to present the somewhat diverse subjects
listed in the program. It was, indeed, a fortunate thing for me that I attended. Seldom have I witnessed a more smoothly organized and effective
conference. Even with your outstanding track record in setting up productive meetings of this type, you must have been impressed with the degree of

participation on the part of all those who attended. Without exception, the speakers were impressive, and each one placed emphasis on the practical aspects of their presentations. On a scale of 1 to 10, I'd give this conference a resounding "10." Thanks for doing such a great job, Dr. Fielding.

(b) I had only one regret concerning your performance at the seminar in Chicago. My entire staff was not with me. It was an exceptionally fine meeting, Bob, and I intend to have several discussions with our people here to fill them in on the topics covered. All of us have attended meetings (both within and outside the organization) and walked away wondering why the end result was not more productive. Not so in this case. I left with a notebook full of "things to do" when I returned to the office. Fortunately, the program brochure indicated that a full transcript would be made available within the next few weeks. I'm not sure who I should direct this request to, but we would like 12 copies of this transcript, the reason being that I want to make a copy available to each member of my staff. If they draw as much value from the written transcript as I did in attending the meeting, all of us will increase our contributions to projects we now have underway.

(c) Just as you were elected to serve as program chairperson at the March meeting, I have been elected by our staff to express our sincere appreciation for the fine job you did. Anyone who is able to capture and *hold* the attention of our group for two straight hours without losing their interest deserves not only my thanks, but my admiration as well. I don't know how you managed to deliver such a fine program on such short notice, but you did—and all of us will benefit from it. Of course, there will inevitably be a penalty you will have to pay for all this. Specifically, the program chairperson has not been designated yet for the annual meeting in October. There is no question in our minds as to who this person should be. Will you accept, Dr. Martin? We would be most pleased if you would, and rest assured we stand ready to offer any assistance that would help you repeat your fine performance of May 4.

● *EXTENDING THANKS FOR PRESENTATION AT*
   CORPORATE MEETING

General Guidelines

The preceding letter expressed appreciation to the chairperson who had overall responsibility for a successful meeting outside your firm. Letter 7-9

is directed toward the individual who delivers particularly significant remarks at an "inhouse" meeting. Often, the perceptive observations offered
at such a meeting can stimulate constructive thought and action toward an
important goal. The presentation itself might be either formal or im-
promptu, but in either case the individual deserves recognition, and thanks,
for his contribution. A brief note is all that is required. It should be written
as soon as possible after the event and include reference to a few specific
reasons why you felt the speaker's remarks were especially helpful.

Letter 7-9

BPI COMMUNICATIONS, INC.
475 Terrace Avenue
Minneapolis, Minnesota 00000

May 15, 19—

Dr. Walter Bradley
Metropolitan Technical Institute
1234 Hope Street
Chicago, Illinois 00000

Dear Dr. Bradley:

Your presentation on "Probing the Future of Fiber Optics" at our
annual corporate manager's meeting was both informative and exceptionally interesting. In fact, your incisive and insightful
remarks turned out to be the highlight of the meeting. We thoroughly enjoyed the "sneak preview" you provided about future
global communications, and the favorable comments following your
presentation reflected the consensus. All of us extend our warm
thanks for an outstanding presentation, Dr. Bradley. I do hope
you'll be able to join us again next year.

Cordially,

Alternate Approaches

(a)   Your remarks concerning current research studies in digital avionics had quite an impact on all those who attended our monthly staff

meeting. We were especially interested in your comments on hardware assessment and validation. Since this is a relatively new field for us, we were also impressed with your approach to systems design and the many ways in which this relates to simplified and more systematic maintenance procedures. The concepts you outlined for fault tolerance will also be of help to us. All in all, Dr. Carey, it was a memorable presentation and one that will benefit each member of our design staff. Thank you very much, and I hope we will have the pleasure (and benefit) of your attendance at a future meeting.

(b) We had not expected you would be able to join us at yesterday's meeting, but I'm so glad you did. And that is not only my opinion. All of our staff personnel join me in expressing appreciation for your support, and for your exceptionally helpful guidance. Quite frankly, we were not familiar with your simplified mathematical calculations used to determine the beam-on time necessary for delivering a prescribed radiation dose. There was also extraordinary interest in your quality assurance program and the way in which this spells out the daily, weekly, monthly, and quarterly tests to perform on each type of equipment. We can understand why you have such a preeminent reputation in the field of radiation therapy. Thank you so much for joining us, Dr. Johnson.

(c) Your extremely interesting discourse on digitizing X rays made an exceptional contribution to our staff, and I want to thank you for this. Although we have, for some time now, been involved in imaging plate and digital fluorography technology, our experience is somewhat limited in the redesign of conventional radiographic equipment. You gave us a fresh insight into this area that will be most helpful as we probe further into this field. Your intriguing comments on a four-step procedure for reconstructing 2-D images from 1-D projections will also be helpful. It seemed remarkable that you were able to cover such a wide spectrum in such a short time, Dr. Friedman. Be assured that it was time well spent. Our staff joins me in expressing our sincere appreciation.

● *CONVEYING APPRECIATION ON SPECIAL OCCASIONS*

General Guidelines

When an important project is successfully completed, or perhaps at the end of an especially busy year, Memo 7-10 represents one way to express your appreciation. Occasions such as these offer opportunities to demonstrate your thoughtfulness both to clients and members of your staff. A brief note conveying a simple but sincere thank you is all that's needed. Regardless of the circumstances, identify the purpose of your remarks and avoid being effusive or turning the communication into a pep talk. The objective is to commend, not to sermonize or issue a plea regarding future performance.

Memo 7-10

December 14, 19—

TO: Staff Members

FR: Gerald Wainwright

While we spend a great deal of time looking ahead toward new products and design techniques, it seems appropriate in the present season to look back for a moment. In doing so, I am reminded once again of the impressive things we've been able to accomplish over the past year. Without question, this progress can be traced directly to the contribution made by each and every member of our staff.

You have my deep appreciation. And with equal sincerity I extend warmest good wishes to you and your families for a memorable holiday season and a very healthy, Happy New Year.

Gerald Wainwright

Alternate Approaches

(a)   Satchell Paige had a point when he said never look back, but at this time of year I can't help but think of the many ways you've been helpful over the past 12 months. There have been countless occasions when you've gone that extra mile to provide data we needed, and we're grateful to you. I realize a service agency such as yours must cope with a wide and diverse group of clients, and there must have been times when you wished we were less insistent in some of our requests. But you never complained (at least to us), and you always managed to deliver top performance at times when we needed it most. Thanks, Tom. It's great doing business with you.

(b) It isn't often that I feel compelled to write a letter of this type. In this case, it is both an obligation and a pleasure for me to do so. Last week when a serious mechanical problem in our computer mainframe brought down all the systems, we called upon Mr. Holroyd, your systems engineer, at his home. And even though it was in the middle of the night, he managed to arrive at our plant within one hour after we called him. After working steadily on the problem for almost five hours, Mr. Holroyd solved the problem, and by 6 A.M. all systems were up and running. You may not be surprised by this because Mr. Holroyd works for you, and you're probably aware of the skill and dedication he puts into his work. But having no previous experience of this type with your organization, we were delighted and even astonished by this performance. I hope you will pass along our sincere appreciation to Mr. Holroyd.

(c) I understand Helen Stanton is a supervisor on your staff. As one of your key people, she probably has a bright future ahead of her. But if she ever decides to become a research specialist, I hope you'll refer her to me: She performed a service for us last week that saved many hours of frustration and wasted effort. Some of the systems we're working on require a considerable amount of demographic research and statistical analysis. One of my people (Doris Kenyon) happened to mention this to Helen at lunch. Two days later Helen delivered to me a detailed analysis of data she compiled from market studies the headquarters office completed earlier this year. We had not even been aware of this, but obviously Helen knew of it. She not only organized the data, but interpreted the results in ways that made it more meaningful to us. I called Helen to thank her, but I wanted you to know how much we appreciate this extra effort on Helen's part.

● *COMPLIMENTING STAFF ON EFFORTS DURING DIFFICULT PERIOD*

General Guidelines

A colleague once remarked that management too often communicated to the troops via the "three E's": *epistles, edicts,* and *exhortations.* A worthwhile antidote to this is presented in Memo 7-11. Following the solution of a protracted problem or any especially difficult period, a thoughtful memo of this type can provide a welcomed change of pace. It not only provides evidence of your level of awareness, but conveys appreciation for the particular effort identified in the memo. As with most of the preceding letters and memos, brevity is a key factor. Another dependable guideline is to keep the focus on the achievement itself, not on "management's view" of the accomplishment. The tone should be warm, friendly, and of course, sincere.

Memo 7-11

January 26, 19—

TO: The Engineering Staff

This personal note expresses my heartiest thanks to each and every one of you for making such a fine effort to be on the job during the brutal three weeks of the transportation strike. Such loyalty and professional dedication on the part of our staff has been the key contributing factor to a successful engineering performance over the years.

With the strike now over and with a most impressive product agenda coming up, I would hope that all of us could regain the momentum we lost and have the best first quarter in our history.

Again, thanks very much.

Alternate Approaches

(a)   There was a time just a few months ago when I thought I'd never be able to write this memo. Those were the dark days, when every conceivable problem seemed to crop up in our development of the X11, including the breakdown of procedures that were supposed to be foolproof. While there were many factors that enabled us to complete the project on schedule and within budgetary limits, I think one in particular made the crucial difference. And that was the "can do" attitude maintained by every member of the staff. This became a "must do" attitude toward the end, and I just want you to know I'm glad to be associated with our staff or professionals. You have every right to be proud of the X11. It came through all the field testing with flying colors, and this memo expresses my sincere appreciation for your efforts.

(b)   While this memo deals with events following the loss of our Worcester plant, it will not cover the factors that led to this loss. All of us have been made aware of these factors by now, so that is not my purpose in writing to you. My primary objective here is to tell you how pleased I am with the way all of you have responded to the extraordinarily difficult period that followed this loss. A reduction of 30 percent in manpower, coupled with the need for extensive repairs to some of our antiquated equipment, made everyone's job a lot more difficult. There were all sorts of reasons for dissatisfaction, grumbling, and low morale. Who prevented these negative qualities from surfacing? You did. It now appears we've just about turned the corner, and I attribute this to the professionals remaining on our staff. Thank you very much for all your fine efforts during this difficult period.

(c)   The merger of two corporations represents a difficult and challenging time for all concerned. It can also represent a period of great opportunity, but only if the personnel involved choose to look at that way. I am writing to tell you how pleased I am, not only with your open-minded attitude toward the merger, but with the individual actions you've taken to make it a success. There has been no lapse in production and very few delays in shipping, the design group has increased its output, and the engineering staff not only completed the L400 prototype, but did so within the cost and time restraints originally imposed. This is impressive. It also provides solid evidence that we are ready, willing, and able to capitalize on the many opportunities opening up because of this merger. To each one of you, my sincere thanks.

● *LETTER TO COLLEAGUE REGARDING FAMILY MEMBER'S ILLNESS*

General Guidelines

The first essential of any letter expressing concern over the illness of a family member is empathy for the reader. Letter 7-12 reflects this element, and does so in a brief, sincere way. Do not burden your colleague with graphic descriptions of your concern. In situations of this type, the recipient simply wants the consolation of your friendship, not a soliloquy. On some occasions the letter will include an offer to help. This does not apply to all situations, but if the intent is sincere and follow-up certain, an offer to help at the end of the letter will be a real comfort to the reader—whether or not the offer is accepted.

Letter 7-12

BENTON MANUFACTURING CORPORATION
880 12th Street
Harborville, Indiana 00000

March 31, 19—

Mr. Ronald Ingram
744 Ames Avenue
Maxwell, Indiana 00000

Dear Ron:

Your friends at the Benton Corporation are very sorry to learn of your wife's illness. Please accept the flowers we have sent along with our sincere wishes she will have a fast and complete recovery. If there's anything at all we can do during this period, Ron, I hope you will let us know.

Sincerely,

Alternate Approaches

(a)   Word of your husband's illness just reached me, and I hope you will extend our very best wishes to him for a fast and complete recovery. We understand he will be in the hospital for several weeks, and we want *you* to understand that our concern for both of you includes our support for the assignment you were working on. Please don't worry about this. Your associates on the staff will see to it that this work is carried on during your absence. There may be other things we can do for you as well. If so, please don't hesitate to call on us. We are not only your professional colleagues, we are your friends.

(b)   Having children of my own, I can sense what a dreadful time this must be for you and your wife. We learned of Tommy's accident this morning, and two thoughts immediately occurred to us. First, of course, was the hope that his progress would continue and that he would be back home with you soon. We also want you to know that not only are our thoughts and prayers with you, but we stand ready to help in other ways if we can. One thing comes to mind, and perhaps you can help us with this. As soon as Tommy is able to receive an expression of our friendship, I hope you will let us know. It's unlikely a youngster of his age would want the traditional flowers, but perhaps there is something we could send that would raise his spirits a bit. Would you give us a suggestion or two? In the meantime, we send our warm regards and very best wishes to you, to your husband, and to Tommy.

(c)   Even though it has been almost a year since I first met your wife, the memory of that meeting is still fresh in my mind. Seldom have I known two people who seemed so ideally suited to each other. I can only imagine how concerned you must be regarding her illness. Please know that you are not alone. All of your friends on the staff share in this concern and extend our most sincere wishes for her early recovery. This is certainly not a time to talk business, and I have no intention of doing so. I just want you to know that everyone here is going to pitch in and handle your assignments in the very best way we can while you are out of the office. Your place is with your wife, and I hope you'll tell her that our thoughts and prayers are with her too. Are there other ways we might be of help, Bill? If there are, I hope you will let us know.

A WORD ON STYLE

Sidestep Grammatical Problems

When a problem in grammar is tricky or obscure—or when the correct choice of a word or phrase sounds wrong—the best solution is often that of the Cowardly Lion. Back off. Get away from it. Instead of puzzling over two poor choices, or taking a hopeful guess, simply forget the sentence as you first drafted it. Rewrite it. That kind of an end run will often take less time than ceaseless deliberation, and it will usually result in a better sentence.

Chapter 8

PRODUCTIVE MEMOS AND LETTERS TO STAFF MEMBERS AND ASSOCIATES

This chapter deals with some of the more particular situations that require written communications. A wide scope is covered, ranging from negative responses to a client's request, to ways in which you can quickly communicate budgetary reductions. As indicated earlier, many letters throughout the book fit neatly into more than one category. Similar situations are dealt with in different ways, and these differences can be important to you. By locating the letter or memo that most closely fits your needs, you can further reduce the amount of adaptation that may be required. Regardless of how a particular letter is categorized, remember to keep the focus on the reader when phrasing your remarks. Wherever possible, take the "You" approach rather than "I" or "We."

While the purpose of each letter or memo will vary, most are designed to establish, and enhance, a feeling of goodwill and cooperation. The element of goodwill is not a one-time thing, but should be designed to continue a productive relationship. This is illustrated by the first letter that follows. Note that while the writer takes a negative approach to the client's request, the suggestion is also made for a realistic and perhaps more favorable alternative, and one that could more satisfactorily fill the client's needs. In each instance, the factors involved are spelled out clearly, and any action or response that may be necessary should be specified with equal clarity. While most communications to staff members are put in memo form, there will be instances where you may feel it more appropriate to use a letter format, sending it to the staff member's home. This might be particularly appropriate, for example, when welcoming a new staff member to your organization. (See Letter 8-10.)

The circumstances and number of key factors to be covered will determine the length of each communication, but every effort should be made to limit your remarks to one page. The type of "inspirational" approach taken in Memo 8-4 may run a bit longer, but situations of this type would be the exception.

● *CONVEYING NEGATIVE RESPONSE TO CLIENT'S REQUEST*

General Guidelines

Diplomacy is a basic requirement whether you are refusing a request or making one. As indicated by Letter 8-1, the first step is to indicate clearly the reasons for a negative response. Whenever possible, this should be followed by suggesting an alternative or two for the recipient to consider. A certain amount of finesse is required. This is particularly true if your response could affect your relationship with an important client. Excessive embellishment should, of course, be avoided. State the facts clearly and concisely; then close with your alternate suggestions and/or your desire to be of greater help at some point in the future.

Letter 8-1

MATSON DESIGN CORPORATION
1214 Wilbur Street
Wentworth, Michigan 00000

April 24, 19—

Ms Jean Evans
450 12th Avenue
Chicago, Illinois 00000

Dear Ms Evans:

I hasten to answer your letter concerning the amplifier circuit boards, because I know time is important to you and your customer. This particular board is completely out of our line, and I think a specialist in this area could be much more helpful to you. By the time we found a source and placed a special order, a good deal of time would be consumed, and the price might also be higher than through the channels you would ordinarily use. If, in spite of this, you'd like us to proceed in solving the peculiar problems this board represents, please send your instructions together with diagrams, any special component requirements,

preferred completion dates, and so on, and we'll make every effort to be of service to you.

Sincerely,

William Smith
Design Engineer

Alternate Approaches

(a)   Having been in a similar predicament a while ago, I appreciate the
dilemma you're faced with. If there was a way I could help you without
compromising our own corporate policies, I'd be delighted to do it. The fact
is, management here has taken a firm, inflexible position on the matter of
making names and addresses of our clients available to anyone outside the
corporation. We have been forbidden to do this, despite the fact we'd really
like to be of help to you. I do have two suggestions that you may want to
consider. First, the latest edition of Standard & Poor's could be of some
assistance. Next, I have enclosed some data on list brokers in your area,
along with comments regarding the ones we've used in the past. I do hope
these alternate courses will prove productive, Mr. Baker.

(b)   You had no way of knowing this, but the request for changes in the
X100 design arrived here at the worst possible time. Our plant shuts down
during the last two weeks in August, and the two designers involved in this
project will be on vacation. If it is vitally important to you that these
changes be made prior to August 24, there is another course of action you
might consider. And if you decide to do this, I'll be glad to help in any way I
can. Specifically, we can subcontract this out to a firm we occasionally use
on a freelance basis. I can vouch for their skill and feel reasonably certain
they would be willing to take on the assignment. I have no way of knowing
what their fee would be, but if you want me to investigate this, I'll be glad to
do so. I want to be of help, Mr. Taylor, so just let me know if you'd like me
to pursue this further.

(c)   Your request for adjustments in the delivery schedule agreed on
earlier has been discussed (at great length) with our suppliers. If it was a
question of our firm being the only one involved, you can bet we'd be glad
to make the changes you requested. Unfortunately, this is not the case. Only
two of the five suppliers would be able to effect the changes requested, and
to proceed on this basis would result in a problem far greater than the one
you have now. Perhaps we could reopen the bidding process, but I hesitate

to do this because I seriously doubt this would be in your best interests. Moreover, it is quite likely this would lead to additional delays and, ultimately, higher costs. Are there other approaches you'd like us to consider, Bob? After you've given this some further thought, please let me know. We really do want to help.

● *RAISING QUESTIONS ON A SUBMITTED BID*

General Guidelines

Little is to be gained by routinely requesting cuts in submitted cost estimates, except a reputation that suppliers will bear in mind when submitting future bids. On those occasions that justify questions or a discussion on cost reduction, Letter 8-2 can be adapted to suit your needs. Some assurance needs to be given that the bid in question is being seriously considered. This should be conveyed in the opening paragraphs. After stating your position clearly and concisely, you may also want to express a desire to see the situation resolved quickly so that additional progress can be made toward a final agreement.

Letter 8-2

KINGSLEY CONTRACTORS, INC.
404 Newport Avenue
Ft. Ames, Indiana 00000

April 25, 19—

Ms Jane Walker
ABC Design Specialists
8942 Main Avenue
Albany, New York 00000

Dear Ms Walker:

I have gone over your sketches and plans for the Westbury Arms with our management and we are pleased with them. There is only

one negative factor. On the basis of competing bids, we feel your
price is somewhat out of line. If you can manage to shave the
price by about 15 percent without sacrificing work quality, I
believe we could come to an agreement, and the assignment would
start almost immediately. Please review your price estimate and
call me within the next day or two about this. If this modest
reduction results in changes in materials, we would have to
have another conference. However, if we can come to an agreement—
and I hope we can—the legal department will then draw up a contract.

Thank you for the effort you put into your proposal, Ms Walker.
I hope we will have an opportunity of working together in the
future.

Sincerely,

Richard Roe
Design Engineer

Alternate Approaches

(a)   Your proposal regarding the hydraulic valve modifications was well
organized and well presented. We appreciate the obvious effort you put into
this. After careful review by appropriate members of our engineering staff,
there appears to be only one aspect that causes concern. If you will review
the layout again (see enclosed diagram), you will notice the change we were
forced to make in dimensions for the enclosures. This would require a
corresponding change in design of the valves; otherwise, we would have a
costly maintenance problem whenever servicing is required. We have some
thoughts on how this might be resolved, but since you would be the primary
supplier, we would much prefer to have the solution come from you. Please
give this further thought and get back to me with your recommendations as
soon as possible. It would still be important to stay within the cost and time
estimates you submitted earlier.

(b)   Of all the bids submitted on the Worthington project, we found
yours to be one of the most interesting. Quite frankly, we were impressed
with your ingenious approach to problems presented by this project. Of
course, there is a natural concern when breaking new ground or using an
especially "innovative" approach toward a project of this type, and I have

an idea that might help us overcome this problem. Specifically, what would you say if we asked you to build a prototype for us? We would probably have to make this a separate project, but I suspect the successful completion of the prototype would help to eliminate any doubts as to whether we should move ahead on a full scale. If you are receptive to the idea, please give me detailed cost figures and a reasonably firm estimate on how long it would take you.

(c)   At this point, I can think of only one factor that would prevent us from signing a contract with you for the voltage regulator ICs. While the cost figures were slightly higher than other bids we received, your willingness to include heat sinks made the difference. This, plus the record you've established for dependability, would normally have caused us to move right ahead to the contract-signing stage. The factor that prevents us from doing so is your reference to estimated delivery time. All our production and marketing plans are built around the introduction of our new RL power supplies in March. There is no way we could do this if the ICs are not delivered to us, at the very latest, by January 15. I realize this would require an exceptional effort on your part, but we're not talking about a one-shot arrangement here. If you can manage to meet this delivery schedule, I feel certain it would be in your long-range best interests. As soon as you've talked with your appropriate people about this, please give me a call.

● *AGREEING TO REVISED SPECIFICATIONS*

General Guidelines

There will always be occasions when original specifications, carefully designed and approved, lead to problems that require some adjustment in the original concept. The difficulty, very often, is in knowing when that point has been reached. Maintaining an awareness of this possibility should not only be standard procedure, but should be encouraged—and acted on at the optimum stage of the project's development. Letter 8-3 covers one such situation, and adaptations can also be drawn from the alternate approaches described. If the changes require any adjustment in an earlier letter of agreement or contract, your attorney should also be consulted.

Letter 8-3

MAXWELL MANUFACTURING CO.
420 Abel Boulevard
Carswell, Maine 00000

April 25, 19—

Mr. Henry Bramhall
Magna Builders, Inc.
3310 Broad Street
Lewiston, Maryland 00000

Dear Mr. Bramhall:

I have examined the bracket samples, and my opinion is that it
will be quite all right to provide #2912 in an aluminum anodized
finish. This would be an acceptable replacement for the earlier
brackets specified in our agreement of March 12. Of course, the
other terms still apply, including a 10-year guarantee on all
parts provided by your firm. Am I correct in assuming that the
guarantee will also apply to the substituted brackets?

Please confirm this before proceeding.

Sincerely,

Alternate Approaches

(a)   I appreciate your notifying me about the change in suppliers for the
cable terminals ordered on March 4. This came as less of a surprise because you alerted me to the possible necessity of doing this in your letter of
January 12. While we agree with your reasoning, I want to underscore a
requirement that is spelled out in our contract. There must be no delay in
the delivery date we agreed on earlier. Your assertion during our telephone conversation that the design, quality standards, and dimensions
would remain the same is reassuring. To make certain of this, we'd like
you to have the new supplier produce 50 units geared to the specs in the
contract. They should first receive your stamp of approval before being
sent to us. We will perform the same tests we had planned to conduct for
units which were to have been manufactured by the original supplier.

(b)   We tend to agree with your suggestions regarding changes in the package designs, but I'd like to reinforce our general agreement with more facts and figures. For example, going to four colors would undoubtedly result in a substantial cost increase. Your letter indicates you agree with this assumption, and now we need to pinpoint the amount of this increase, geared to the quantity figures indicated earlier. Also, it would seem that some consumer studies are in order to see whether the changes suggested are justified. I suspect this will confirm our joint thinking on this, but we need to act on the basis of facts, not suspicions. As a first step, I suggest you consider running a small-scale test on the new adhesive container designs and score them against our present package. How quickly would you be able to move ahead on this, and when do you think you'd have the results?

(c)   Our quality control people have examined the signal amplifiers you left with me, and we have decided that a VHF gain of 25 dB is acceptable. I understand these were the first units manufactured during a test run and that is the reason our logo did not appear on each unit. All other terms of our contract still apply, and this letter should be considered as authorization only for the change in the VHF decibel gain rating. I have enclosed two copies of the amended agreement. Would you please initial the paragraph indicated and return one copy to me. We should also emphasize that there has been no change in the requirement for a UHF gain of 20 dB. This, along with all other provisions of the original contract (with the exception of paragraph L), is still in effect.

• *ENCOURAGING INNOVATIVE APPROACHES BY SUPERVISORY STAFF*

General Guidelines

In addition to the normal flow of communications to staff members, there are times, as illustrated by Memo 8-4, when an "inspirational" message can be beneficial. This is not as easy as it might sound, and viewing it in a simplistic way too often results in an ego trip for the writer. The purpose is to inspire, not to lecture. It should be viewed as an opportunity to encourage a more positive, constructive attitude toward staff management, as opposed to specific instructions related to a particular project. The chain reaction that can result will frequently contribute to projects underway, but again, the primary objective is to promote more innovative "can do" thinking among key staff members.

Memo 8-4

January 15, 19—

TO: Engineering Supervisors

FR: John Maxwell, Director

Technical supervision has been described (rather paradoxically) as more of an art than a science, and our successes can seem fleeting while our failures are glaring, obvious, and immediate. My own outlook has always been that a good supervisor's reach will sometimes exceed his grasp.

If the supervisor's efforts, despite careful preparation, intelligent conception, and solid execution do not meet with hoped-for success, something is still gained through the experience. In the process of stretching, growth inevitably takes place for both the supervisor and the supervised. Without that extra stretch, not only is our personal and professional growth arrested, we are almost inevitably consigned to a fate of vegetation and failure.

Consider for a moment the examples of our colleagues who have just recently risked the prospect of failure, whose reach stretched them to the utmost. Without their courage and adventurousness, we would not have been granted a $175,000 Title VII program, would not have become involved in a cooperative effort with the Greenville Technical Institute for engineering graduates, would not have been given a grant from the Coordinating Council for Scientific Advance, and would not have mounted an extraordinary effort that led to delivery of the XC-400 project one week ahead of schedule.

The capacity to take considered risks, to face up to difficult and complex issues, will help us meet the challenges represented by new products now in the development stage. In some instances we may fail, in others we may achieve only partial success, but at least we will not be prisoners of cynicism or hostages to despair. The rate at which we are now progressing convinces me that our successes will far outnumber our failures. I hope you agree.

Alternate Approaches

(a)   On the basis of discussions at our recent meetings, I sense that most of you agree we need to take a fresh look at some of our departmental procedures. Many of the procedural changes necessary have been identified. Others will undoubtedly come to light as we proceed to make adjustments in practices that are no longer as effective as they should be. The attached list summarizes the changes we have agreed on to date. But the principal reason for this memo has to do with a more basic adjustment. That is, the need to place greater emphasis on more qualified, and more productive research. Too often, and the evidence supports this, we have proceeded on important projects with insufficient data, and on other projects with data that later prove to be partially or wholly incorrect. This requires an attitudinal change, as much as it requires procedural changes.

(b)   First, I want to express my appreciation to the supervisory staff for the way you managed to maintain production standards during the recent labor difficulties. It was not easy, and even though a settlement has been reached, I think we'll continue to feel the repercussions for some time to come. All by way of saying that I think all of us learned something from this experience. It seems to me that the most important thing we have learned is something so basic that one wonders why we lost sight of it in the first place. And that, quite simply, is the need to listen. I realize it isn't always easy, under the pressure of deadlines, machine breakdowns, meeting quotas, and so on. But all of us have experienced the frustration of talking to someone who doesn't really hear what we're saying. We need to listen not only to what staff members say (and sometimes to what they don't say), but also understand the implication of what they are saying.

(c)   The annual performance ratings have been completed for each staff member, and on the whole, I believe they reflect a good deal of individual progress over the past year. From the standpoint of manpower, it appears we are now in better shape than we've been in for a long time. This is good news indeed. The one point of major concern represents a potential problem that I think all of us should be aware of. The concern stems from remarks made by several staff members during the rating sessions, specifically, their feeling that management is locked into our current policies and procedures and really does not want ideas or suggestions for improvements. I'm not certain as to how widespread this feeling is, but if it exists on the part of only one staff member, we should be concerned about it. Please let me have your ideas on how we can provide greater incentive and encouragement to those employees who have suggestions for improvements.

● *CONVEYING, OR DENYING, SUPPORT TO*
  OUTSIDE ORGANIZATION

General Guidelines

Those involved in technical professions are frequently called on to support activities sponsored by others in the community. The activity may or may not have a direct relationship to your organization. It goes with the professional's "territory," however, to be involved with some of these efforts, either on a personal basis or as a representative of your firm. Almost always, such activities involve writing letters. This involves an added drain on your time, but the letters are important and frequently deserve more attention than they get. Letter 8-5 and the alternate approaches will help simplify your responses to requests of this type.

Letter 8-5

BENTLEY TECHNICAL INSTITUTE
1240 Woodward Avenue
Raleigh, North Carolina 00000

OFFICE OF
THE SUPERINTENDENT

January 19, 19—

Mr. Jack Hanna
Instructional Television and Radio
2712 Millwood Road
Columbia, South Carolina 00000

Dear Mr. Hanna:

This letter is to inform you of our interest in the science-related education project described in your letter. We support your application for a block grant being submitted through the

U.S. Department of Education and will make a special effort to have one or more engineering representatives present at the training session that results from the award of this grant.

As I understand it, the venture will be a joint effort between the states of North Carolina and South Carolina, with the primary goal of implementing new science-related education materials into existing curricula.

Please understand that this letter of support does not represent a financial commitment on the part of this Institute. We will, however, be glad to explore other ways in which we may be able to establish a productive working relationship with your appropriate staff members.

Sincerely,

J. F. Hall
Superintendent

Alternate Approaches

(a)   We are pleased and honored to join others in our community who are participating in the fund drive for expansion of the Warwick Regional Hospital. In response to your recent letter, Dr. Willard Randall of our staff will be glad to speak on the topics you suggested, and both of us would like to meet with you prior to the dinner on March 4. Recent developments in pharmacology have provided the basis for an interesting slide presentation developed by our staff, and you might want to include this in the program too. In short, we want to support the hospital's expansion plan and will be glad to lend our assistance to this worthwhile endeavor.

(b)   Thank you very much for thinking of us in connection with the Third International Science Fair, to be held on May 12. We have carefully considered your request for a contribution to the Fair and agree with the global objectives of this fine organization. Although we'd like to respond affirmatively to all the appeals we receive, there is no escaping the fact our funds are limited. Moreover, since our operations are totally domestic in nature, it is our standard policy to concentrate support activities in the areas of local community chests, United Fund drives, and aid for education, particularly in localities where we have our major facilities. I am sorry we cannot contribute to the Fair, but we do wish you every success with this worthy project.

(c)   Your outstanding work on behalf of senior citizens in our community is well known, and I feel honored by the invitation in your letter of May 2. Under different circumstances I would welcome the opportunity to become a member of your Committee. You may be aware, however, that I am deeply involved in a number of community activities at the moment. If I were to take on additional responsibilities, coupled with the considerable amount of travel my job requires, I couldn't possibly do a good job in either community project. It seems likely that you have also faced this problem in the past, and because of this, I hope you will understand the position I'm in. Nevertheless, I am grateful you thought of me, and if there's any other way I might be of assistance, I hope you'll let me know. We could, for example, publicize your committee's fine work in our corporate newsletter. If you feel this would be helpful, just send me a transcript of what you'd like us to include, and I'll see that it appears in the next issue.

●   *REJECTING PROPOSAL FROM VENDOR OR CLIENT*

General Guidelines

The type of communication illustrated by Letter 8-6 represents not only courtesy on the part of the writer, it has its practical aspects too. Formal bids and other types of proposals are not always limited to singular events. You may reject a particular proposal, but see something in it that suggests a contact should be made at some future date. Even if this is not the case, rejecting someone's proposal courteously and thoughtfully is the type of action that can eventually pay dividends even if they are not anticipated. You need not expand at great length on the reasoning behind your decision. Just a brief note, phrased in a way that will not alienate the reader, is all that is required.

Letter 8-6

K. B. STEEL CORPORATION
346 Carroll Street
Brooklyn, New York 00000

March 26, 19—

Mr. Richard Kaye
Glocester Corporation
Albany Road
Carwell, Pennsylvania 00000

Dear Mr. Kaye:

Thank you for the time and effort you spent in preparing your proposal of January 16.

We regret that we are awarding this contract to another vendor because of price and delivery time considerations.

Your proposal was well organized and well documented, and you may be assured that your firm will be included in any future invitations to bid for electronic controls needed by our organization.

Sincerely,

Alternate Approaches

(a)   We appreciate the obvious time and effort you put into your proposal of May 12. This made the final choice an extremely difficult one because consideration was narrowed down to your proposal and one other. We have decided to issue a contract to the other firm, primarily for a reason that was beyond your control: specifically, the geographical location of the other firm, which led to a correspondingly shorter delivery time. There may, however, be future occasions when the delivery factor may not be as crucial as it was in this instance. Provided a situation of this type develops, you may be sure we will be in touch with you again. Thank you for submitting your proposal to us, Mr. Kenyon, and for your patience in waiting for this response.

(b)   We do not often receive proposals from clients similar to the one contained in your letter of June 4, and I want you to know we appreciate the thought behind your proposal. After giving the matter a good deal of thought, and discussing it with our appropriate people, we've decided not to add to our consulting staff at the present time. The decision has no relationship whatever to the considerable experience you've had in this field. Your credentials are excellent. It is simply a matter of timing—and the limited number of contractual arrangements we now have underway that require the skills your firm could offer. Naturally, we are hoping that the day will soon arrive when we need to expand this part of our operations. When this point is reached, Mr. Garvey, I hope your firm will still be interested. Until then, thank you very much for your letter.

(c)   While we have decided to accept one of the other bids submitted for construction of the Apex facilities, I feel compelled to send this note to you because of our past relationship. If it is any consolation, our engineering staff definitely wants your firm to be considered for future assignments. The decision in this instance was based on both price and delivery constraints. As you know, this contract is simply the first step in an ongoing and rather extensive building program. At appropriate points along the way, and as we enter each phase of the operation, your firm will be considered when requesting additional bids from electrical contractors. In the meantime, thank you for your bid on the Apex project and for the interesting suggestions you offered during our recent luncheon.

● RESPONDING TO REQUEST FOR ADVICE

General Guidelines

A request for your counsel or opinion on a matter of importance can be a challenge to your skill at diplomacy. This is true whether you offer the advice, or decide to refrain from doing so. Unless the action is merely to assuage the giver's ego, the advice must be skillfully and thoughtfully given. The advisor, unless he or she is very careful, may appear to be placing himself or herself in an exalted position. Your suggestions may also conflict with the course of action the recipient really wants to take. When you decide to decline, Letter 8-7 can easily be adapted to your situation. The alternate approaches listed will be helpful in the event you decide to respond affirmatively.

Letter 8-7

JEROME KENT ASSOCIATES
22 House Street
Hampton, Virginia 00000

April 25, 19—

Mr. J. A. Smith
Livingston Corporation
Benson, Missouri 00000

Dear Mr. Smith:

As you say, there are several ways you can go to solve the problem described in your letter. However, in all frankness, I cannot
and should not take the responsibility of advising you. For one
thing, I do not know nearly enough about the inner workings of
your company. If the possibility has not already occurred to you,
I do have one suggestion. Specifically, that you call in a team
of professional engineering consultants who have the experience
and expertise necessary to work with you on this. Even then, of
course, the final decision will be up to you.

In any event, I want you to know that you have my sincere best
wishes for a successful solution to this problem.

Cordially,

Alternate Approaches

(a) I'm flattered that you want my opinion on whether you should
continue with the Gladstone project. It might be considered somewhat
adventuresome to move ahead on this with so little hard data available, but
I believe the old saying that nothing ventured, nothing gained, might apply
here. I cannot tell you what to do, but I can tell you what I would do. And
that, quite simply, is to move ahead only to completion of the blueprint
stage. This would enable you to develop a more reliable judgment on the
question of whether the eventual result will be cost effective or not. If the
hard facts clearly indicate the necessary profit margin would be achieved,

fine. Provided the facts support the opposite conclusion, then I would drop the project and move on toward improvement of your present system.

(b) In response to your request, I have gone over your organizational plans for the new warehouse unit. Frankly, you exceeded my expectations in the sense that I could not find one major aspect that was overlooked. One thought did occur to me, however, and because we have had some recent experience with hydraulic lifts, I want to share it with you. Because of the endless maintenance problems we had last year with the Stanton units, we finally decided to switch all our hydraulic equipment to the Smithson line. I have no way of knowing why you decided on Stanton, and quite possibly that was the wisest choice you could make, considering any budget restraints you might have been faced with. Perhaps, in all fairness, it is the best choice from other standpoints as well. It might simply have been an aberration that led to our maintenance difficulties, but this aspect is something you may want to investigate further with other manufacturers that use the Stanton equipment.

(c) You may not realize it now, but I'm going to do you a favor by not advising you on the financing dilemma you wrote me about. While I sincerely appreciate the thought behind your request for advice, the fact is I could do you more harm than good by trying to counsel you on this. It may not generally be known (at least, I hope it isn't), but my financial skills are limited to the selection of people who *do* have those skills. And, in this connection, I have a suggestion you may want to consider. Bob Keynes is one of the sharpest professionals I know in the field of raising venture capital. He's a man of great integrity, and after stressing the confidential nature of our conversation, I talked to him about the problem you described. He is willing to help, but only if you initiate the contact. Whether you get in touch with Bob or not is up to you, but since you asked for my advice, I am suggesting you consider drawing on the skills and experience of a topflight professional in this field.

● *ACKNOWLEDGING COLLEAGUE'S REFERRAL OF JOB*
  APPLICANT

General Guidelines

While the circumstances will vary, Letter 8-8 illustrates one way to respond when an applicant is referred to you by a staff member or other contact. There will be times when such a referral requires delicate handling, for example, when you do not agree with your colleague's opinion of the applicant's qualifications. Nevertheless, a courteous response is required, along with a reasonable explanation of the reason why you have (or have not) decided to interview, or hire, the individual recommended. Regardless of whether you have a positive or negative reaction, an expression of thanks is called for.

Letter 8-8

ACME CONTROLS, INC.
222 Benson Street
Rogers, Oklahoma 00000

April 25, 19—

Mr. Thomas Deveraux
Abco Products Company
2442 Jones Road
Kent, Oregon 00000

Dear Tom:

I will be glad to accept your offer to refer Ms Jane Smith to me for an interview. We're impressed with the biographical data you sent, and if she can cope with our complicated terminology here, we may offer her the job of technical assistant. We're always on the lookout for good trainees, and your recommendation came at an ideal time. Just have Ms Smith call my secretary for an appointment. We'll take it from there.

Thanks very much, Tom.

With best regards,

Alternate Approaches

(a)   Your letter of May 12 was timely. We are thinking of adding to our personnel in the Washington office, and I found your description of Mr. Blake's experience quite interesting. Although our recent advertisement in several trade publications resulted in four applicants we're considering seriously, there's no reason why we shouldn't also interview Mr. Blake. The resume you sent reflected a strong background in systems work, but only limited experience in our field. Nevertheless, I appreciate your thoughtfulness in writing to me about his availability and would like to pursue this further. If you will ask him to give my secretary a call so we can set up a convenient appointment time, we'll proceed with the interview as soon as possible. Thanks again for thinking of us, John.

(b)   It was good of you to remember our conversation of a few weeks ago, when I mentioned the job opening we have in the Worcester plant. The thought occurs to me now that I should have given you a more detailed picture of the type of person we're looking for. In addition to electrical engineering and a strong background in circuitry design, the person we finally select must have an established track record in executive management, plus experience in working successfully with government agencies. This is a pretty tall order, and one that I doubt Mr. Gaynor could fill. Obviously, I must have given you the impression the job was on a slightly lower level. This may have been true at the time, but since our conversation, we've decided to upgrade the position, basically because the current director of this operation is due to retire next year and the individual we hire now should be capable of moving into that position. We'll certainly be glad to consider Mr. Gaynor for other possibilities, and many thanks for referring him to us.

(c)   I suppose this could be considered a "good news/bad news" response to your nice letter of June 16. The bad news is that the position we had opened was filled last week. The good news is that we were so impressed with your description of John Russo that we'd like to talk to him anyway. I realize this could be an imposition on Mr. Russo, but I think you know me well enough to know I would not want to waste his time, or mine. It's just that we are expanding in several divisions of the company, and if he manages to impress our chief engineer as much as your endorsement of him has impressed me, then we just might be able to find a good opportunity for him here. Would you be kind enough to serve as the middle man in this situation, Bill? If so, just tell Mr. Russo what I've told you, and if he'd like to follow through on this, he can give me a call next week. We'll set up an appointment, and, of course, I'll be glad to let you know what happens. Thanks very much, Bill.

● *RESOLVING QUESTIONS CONCERNING
   DEFECTIVE PRODUCT*

General Guidelines

Memo 8-9 covers one of the many situations that relate to defects in the design and manufacture of various products. The corrective action required will, of course, depend on the circumstances leading to the defect. In most cases, however, the situation should be documented and corrective action clearly spelled out, not only for the record, but also for the benefit of those responsible for taking the action. Decisions on the steps to be taken should not be arbitrary, and appropriate members of the staff should be consulted before the issuance of written instructions. Ideally, those closest to the problem should, by their recommendations, form the basis of your memo. Once agreed on and approved, you will want to make certain that all those responsible for the corrective action are fully informed.

Memo 8-9

April 25, 19—

TO: A. B. Wainright, Parts Engineering

FR: Clark Roe, Manager

RE: 16-234792-966 Wafer Pack

I can understand your concern about accepting the reworked wafer packs and am inclined to agree with your opinion, although taking this position would cost us greatly in delay. It occurs to me that there may be a compromise solution.

I suggest you have Triangle reseal ten of the wafer packs in the manner they suggest and deliver them to Quality Control. Let Q.C. then have them assembled in simulated operational status and have the finished product subjected to an operational test. If the wafer packs pass this retesting, I believe it will be safe for us to accept the entire lot of reworked wafer packs, again, subject to all standard tests.

If Triangle will not agree to this sample testing, we have no choice but to wait for the entire 200 to be replaced, and I believe you should stand firm on this.

Clark Roe

Alternate Approaches

(a) You are aware of the occasional problems we've had lately with stamping machines in the Walliston plant. Because of the increasing number of complaints we're getting from the Client Relations Department, we can no longer deal with this by coming up with temporary solutions. This memo is going to all staff members involved, and we need your best thinking on ways to resolve the problem. I have set up a meeting with managers of the Production, Quality Control, and Client Relations departments for 3 P.M. on Friday of this week. I would like each one who receives this memo to give your respective supervisor a written list of your suggestions and to do so no later than Friday morning. You are the ones closest to the problem, and we'll welcome your thoughts on how it can best be resolved. Thanks very much.

(b) I have attached a copy of the year-end figures we've been waiting for, and you have a right to be proud of the increase in both production and sales for our 75-ohm coaxial cable. That's the good news. The bad news is that the cartons we're using for the 100-foot lengths are not sturdy enough, and our distributors are complaining about shipping damages. They are not the only ones. Our Shipping Department wants a stronger carton too. Consequently, I have assigned Bob Wellington to quarterback the development of an improved design, and he will be in touch with each of you about this. I'd appreciate it if you'd give him your ideas and complete support. We need to move on this quickly. In terms of specific objectives, we want the final design approved no later than March 1, production to begin no later than March 15, and full production quota achieved and maintained no later than the end of March.

(c) Stating the obvious is something most people like to avoid, and I am no exception. Having said this, I will proceed to do just that. The primary function of our Quality Control Department is to spot defects, not correct them. I feel compelled to mention this because there seems to be an adversarial relationship developing between our two departments, and that could have a harmful effect on all of us. Please carefully review the attached data covering last month's production run on the X100 units. Note the percentage of defective units, as compared with three months and six months ago. Obviously, the trend is going in the wrong direction, and we need to determine the cause for this—and do it quickly. I have asked the quality control manager to meet with us at our next staff meeting. He will provide a more detailed analysis of the defects, and I'd like each of you to contribute ideas and suggestions at this meeting.

● *WELCOMING NEW MEMBER TO TECHNICAL STAFF*

General Guidelines

Letter 8-10 provides an effective way to welcome a newcomer to your organization. Letters of this type can create a feeling of goodwill, enhance the corporate image, and convey your active interest in the staff member's future with your firm. It is best to put this in letter form, sending it to the individual's home. Personalize your remarks by making reference to something that could only apply to the person you're writing to. There is seldom any need to go beyond one page in length, and whenever appropriate, send a copy to the employee's supervisor. Enclose any additional literature that will help new employees get off to a better start, if they have not already received material of this type. This could be a policy manual, employee newsletters, and/or a copy of the annual report.

Letter 8-10

BROWN, KING & WOOD, INC.
446 Oak Street
Nashville, Tennessee 00000

February 17, 19—

Mr. Matthew Gerhard, P.E.
1520 Oak Street
Nashville, Tennessee 00000

Dear Mr. Gerhard:

I want to welcome you as a new member of Brown, King & Wood, Inc. You can be certain we'll do our best to fulfill our obligations to you as a new member of our organization.

Since any corporation is only as good as its people, we believe that our capable engineering staff will continue to make an ever-increasing contribution toward the overall success of this firm. We hope you will become a key factor in the expansion plans being

developed for the future and look forward to a long and mutually
pleasant association between you and our organization.

Enclosed you will find a copy of our latest annual report and a
separate quarterly report on the engineering staff performance
and objectives. If you have comments, suggestions, or questions
regarding any of this material, I will be pleased to discuss them
with you.

Sincerely,

Charles Brown
Director

Alternate Approaches

(a)   It is a pleasure to extend this greeting to you as a new member of the
engineering staff. You have met several members of our organization during the interviewing process, and I plan to visit your department to extend a
personal note of welcome to you in the next week or so. Actually, the fact
you are a relative newcomer gives you a certain advantage over other
members of our staff. You will be able to view many of the projects underway in a fresh light, and some of your first impressions could be helpful. Do
not hesitate to share these impressions with your supervisor. The great
majority of our existing products originated in the department you are now
part of, and we look forward to your contributions toward future products.
Welcome aboard, Mr. Bradley.

(b)   You are now a member of a fine group of professionals, operating as
a team and with the objective of providing top-quality electronic components for a rapidly growing market. As you become more familiar with your
assignment and better acquainted with other staff members, you will discover how individual contributions have led to the success of our organization. The collective and impressive output of this group has resulted from
these individual contributions, and we look forward to the ways in which
you will add to this effort. Your excellent background with the Apex Corporation will stand you in good stead here, and we were impressed with your
innovative approaches in the field of operational amplifiers. My warmest
good wishes to you, Mr. Bryan, as you begin what I hope will be a long and
productive career with our organization.

(c)   Your decision to accept our employment offer will, I sincerely hope, turn out to be a wise and rewarding decision for all concerned. We look forward to having you with us, and expect to see you on June 1. In the meantime, you will find it helpful to carefully review the material that accompanies this letter. You may also want to share portions of this material with your wife, who will also be interested in the provisions that describe our employee benefit program. As your "partner," we'd like her to be pleased with your new position too. In terms of your initial assignment, we're delighted you had extra courses in this subject area while continuing your education at night and will welcome your solutions to the problems discussed during our last meeting. Once again, welcome to our staff, Mr. MacDonald.

● *EXPLAINING CHANGES IN CORPORATE STRUCTURE OR POLICIES*

General Guidelines

Fundamental and far-reaching changes within the organization often involve complications that affect every member of the staff. Memo 8-11 covers one such situation, and others are treated within the alternates provided. In each case, staff members should be given the facts in a clear, concise way and told how the changes will affect them. Whenever possible, inject an optimistic tone in your remarks and don't be pedantic when providing the details. Your objective is to enlist staff support for the changes, and this becomes unlikely if you cloud the issue with vague or excessive language. They will also expect a clear indication that you are in accord with the particular change described.

Memo 8-11

May 28, 19—

TO: The Engineering Staff

FR: Robert Austin, Chief Engineer

Since April, when I first wrote to you about our proposed merger with Lafayette Software Company, several important steps have

been taken. The final decision came on May 8 when the stockhold-
ers of both companies overwhelmingly approved the plan to merge.

Consequently, I'm happy to announce that the merger between our
organization and the Lafayette Software Company has been final-
ized, and the name of the new company will be THE LAFAYETTE MICRO
COMPANY. The new firm will combine the skill of our hardware engineers with the expertise of their software systems engineers,
effecting an outstanding blend of computer knowledge geared to
small-business and home usage. These are burgeoning fields where
success awaits the right kind of computer systems.

Insofar as our staff is concerned, the merger should mean greater
job security and increased opportunities for all of us. I hope
you are as pleased with this merger as I am. It will enable us to
expand our engineering horizons and make our firm more productive and more profitable in every respect. Any questions or
concerns you have will be fully discussed at our regular staff
meeting, to be held on Tuesday of next week.

Alternate Approaches

(a)   This memo will acquaint you with a change in one facet of our
employee benefit program. You are aware that our scholarship aid program
has been in effect for the past several years, and the record indicates many
staff members, along with the company itself, have benefited from the
program. Effective July 1, a change is being introduced that will affect only
a relatively small number of employees, specifically, those employees who
have requested the company subsidize their interests in nonjob-related activities. Effective July 1, all courses taken must have a direct relationship to
the staff member's present position or to a position of advancement within
the organization. Applications must be approved by your supervisor in
addition to the usual approval from the Human Resources Department. We
are hoping this will result in an increase, not a decrease, in the number of
applications received.

(b)   We are pleased to announce an important addition to our list of
employee benefits. In light of the potentially successful projects now underway, this additional benefit is introduced at a particularly appropriate time.
The Apex stock purchase plan for all our employees will go into effect on
May 1. The details are outlined in the attached booklet, and we urge you to
review this material most carefully. The plan will provide you with a means
of acquiring stock in your company on a regular basis, at a special employee

price, and, if you wish, via payroll deductions. This will provide you with an additional means of building financial security into your future. We hope you will be as pleased with this plan as we are in making it available to you. After reviewing the enclosed information if you have any questions, John Gabacia in the Personnel Department will be glad to answer them for you.

(c) First, let me thank all those who offered constructive suggestions and their support to the research staff reorganization. I know this has been a matter of concern to all of us, and the fact we now have a strong, viable plan is a tribute to all of you. Next, I want to reassure each staff member on the point of greatest concern. This reorganization will *not* result in any layoffs or reassignments for members of the three departments being merged. The purpose all along has been to coordinate the efforts of the three units more effectively, to avoid duplication of effort while increasing opportunities within the departments affected. The attached organization chart summarizes the new structure, and we will have an opportunity to discuss this in greater detail at our next meeting on May 14. In the meantime, if there are urgent matters you want to discuss prior to that date, please call me on Ext. 2457.

● *INFORMING STAFF OF BUDGETARY REDUCTIONS*

General Guidelines

As the saying goes, bad news travels fast. This is one of the primary reasons why it should be conveyed with clarity and understanding, avoiding at all costs unfounded rumors that spring from poor or inadequate communications. Memo 8-12 can quickly be adapted to fit one of the more distressing situations confronting managers and employees alike: a reduction in staff. Budgetary cutbacks, for whatever the reason, need to be conveyed with both sensitivity and candor, a combination that is difficult to embody in a single communication. Decisions to defer the purchase of needed equipment can also have a negative effect on morale and production, and this is covered in the alternate situations that follow.

Memo 8-12

March 28, 19—

TO: William Mullen

FR: Richard Thornton

This is a follow-up memo to our conversation of this morning. As
both of us have become aware during the past six months, economic
conditions have hurt us badly, forcing a reduction of staff in a
number of departments. This is most unfortunate, but we have been
unable to come up with any other viable alternative. Most regrettably, your position is one of those to be eliminated. As I told
you, a lot of hard thinking and long discussions with your supervisor, Arnold Thomas, preceded this decision.

We hope that as economic conditions improve, we will be able to
consider you for another position as soon as one becomes available. Every sincere and good wish is extended to you in locating
a new position. Even though you were with us for a relatively
short time, we extend our thanks and appreciation for the good
work you've done. You may be certain a most favorable endorsement
of your work skills will be passed along to any prospective
employer who contacts us.

Alternate Approaches

(a)   Some of you have probably anticipated the decision communicated
by this memo, but that doesn't make it any easier to write. The fact is, our
budgetary situation forces us to delay purchase of laboratory equipment
itemized on the attached list. I realize this will set off a chain reaction in
terms of its effect on several projects we are currently involved with, but
unfortunately, we have no other choice. The figures for the last quarter
were, as you know, considerably under projected income estimated at the
beginning of this fiscal year. Intensified competition is one of the reasons,
and others will be covered at our next manager's meeting on June 2. In the
meantime, I would like each manager to give me an evaluation of how this
decision will affect your operation over the next three months.

(b)   While I regret the need to discontinue Model X100 as much as you do, the decision was relatively easy to make. Sales have been going downhill for almost a year, and all of us were well aware of this product's shortcomings. Inevitably, this will have its effect on several aspects of our operation, but there is one area I am most concerned about—and I need your very best effort on this. Specifically, our manpower. My estimate is that four staff members will be directly affected by this action. *How* they will be affected depends to a large extent on whether we will be able to bring about appropriate transfers for these people. Their personnel files are attached, and I would like you to go over them very carefully. After doing so, please let me have your recommendations as to what transfer opportunities we should explore.

(c)   I have attached a copy of the annual report to this memo and would like you to review it in detail. There is a message here that, in a sense, provides us with an advance warning of actions that may need to be taken. To put it more bluntly, a downward adjustment of our departmental budget appears inevitable if the first quarter of this year is not in line with projections made earlier. Whether actual results show an increase or decrease, we need to develop plans for either eventuality. First, I would like your monthly report for January to include a projection of total units to be completed by March 31. Next, please let me have a list of reductions in expenses you will implement if first quarter figures are *not* in accord with budget requirements. The attached data will be of some assistance to you in developing your response to both these requests.

A WORD ON STYLE

Aim for Short Paragraphs

Paragraphs in dissertations, technical papers, and the like are often long and thoroughly developed. Because of this, it is hardly surprising that engineers sometimes write longer paragraphs than their readers would prefer. This is particularly true when the subject doesn't involve highly intricate, technical matters. Too much unrelieved text, without any visual breaks, will bother the average reader. As editors of tabloids discovered a long time ago, short paragraphs invite readers, long paragraphs do not.

This is not to say that you, as a professional in a technical profession, should try to emulate the *National Enquirer*. It is only to suggest that long paragraphs can often be effectively broken into two or more paragraphs, with gains in clarity, visual appeal, and effectiveness.

Chapter 9

MODEL LETTERS AND MEMOS OF ACCEPTANCE OR REFUSAL

Whether your letter or memo communicates acceptance or refusal, the common denominator is courtesy. Making your meaning crystal clear, whether delicately or bluntly, as the situation requires, is also important in communications of this type. Some letters require especially careful wording to avoid legal entanglements. In those instances where you suspect there may be a legalistic point to be concerned about, you are advised to get expert advice. It is much better to be counseled ahead of time than to make an error and then seek to be extricated from the resulting tangle. As an example, the writer of Letter 9-1 made certain his position was in accordance with the "terms of our contract" before sending it. Responding affirmatively to a request is a relatively simple matter. It is much more difficult to refuse or say "no." A client who receives a tactless "no" may not complain openly, but may simply take his or her business elsewhere. One way to begin a letter of refusal is to, wherever possible, agree with the reader on some point. This tends to establish a feeling of working together as opposed to an "I'm-right-you're-wrong" confrontation. An alternative approach would be to thank the writer in your opening statement for being candid, for directing the inquiry to your attention, or for some other reason that establishes a tactful, courteous prelude to your response.

Expressing the desire to help, and your regret at not being able to do so, represents a third possible opening approach, the message being that "we would like very much to do what you ask, but the following reasons prevent us from doing so." After letting the reader know you have carefully reviewed the request, and considered it from his or her viewpoint, convey the specific reason for your refusal. Ideally, you will explain the refusal before stating it in concise terms. Try also to end your letter or memo on a positive or upbeat note. If there are alternative courses you can suggest, offer them as an indication that you do, indeed, want to help. Letters of acceptance should be equally specific regarding any action that needs to be taken as a result of your positive response.

● *REFUSING CONTRACTOR REQUESTS FOR CHANGES*

General Guidelines

Letter 9-1 followed the writer's consultation with legal staff on his interpretation of the contract terms in effect. While his position is sound from a legal standpoint, this does not prevent him from expressing regret at this turn of events, and he does so in the opening sentence. In other situations the writer might have softened his stand somewhat, by not referring to the possibility of bringing "the legal staff into the picture." In this instance, however, the circumstances were such that it was felt advisable to do so. Nevertheless, the tone is straightforward and courteous, with the writer's position spelled out clearly and concisely.

Letter 9-1

WYLIE MANUFACTURING COMPANY, INC.
1240 Manley Boulevard
Topeka, Kansas 00000

April 25, 19—

Mr. Richard Caruso
Able Contractors
9012 Busy St.
Allison, Kansas 00000

Dear Mr. Caruso:

I am sorry that your expenses are running beyond expectations on the warehouse extension you are building for us at 17 Robinson Way. Nevertheless, we feel this is your responsibility according to the terms of our contract. Our funds were budgeted in accordance with the figures you gave us, and we are in no position to allot additional money to this project. If this becomes a big issue, and I certainly hope it will not, we will have no recourse but to bring the legal staff into the picture. After you have

completed the foundation, please call me so we can discuss this
further.

Cordially,

Alternate Approaches

(a)   I had hoped it would not be necessary for you to send your letter of
May 12. You will recall our position was conveyed to you during the meeting
we had with your staff on May 10. The changes in bracket specifications you
are requesting were carefully reviewed by our engineering staff, and I regret
we cannot go along with you on this. In order to change the configuration at
this point, we would be faced not only with additional expenses, but the
retooling involved would force us to go beyond the completion date agreed
on in our contract. This, in turn, would set off a chain reaction affecting our
clients and, ultimately, the relationship we hope to maintain with your firm
in the future. This is not an arbitrary decision. It is simply in accord with the
written agreement we have with you, and I sincerely hope you will follow
through on delivering the brackets as specified in our contract.

(b)   Thanks very much for your candid letter of March 4. My response
will be equally candid, and I wish it could be more amenable to your
request. The fact is, however, that the change in specifications you're
requesting would not be in your best interests or ours. We have committed
our firm to delivery of the M30 units no later than April 20, based on the
design specifications clearly specified in our contract. I have called our
clients regarding the changes you described, and without exception, they
were strongly opposed to these changes. If we were to lose even one of
these clients because of the adjustments described in your letter, both our
firms would suffer. As a result, I feel compelled to make it crystal clear
that changes in the original specifications cannot be permitted.

(c)   To give you some idea as to how carefully we reviewed your request,
I had two of our engineers fabricate a prototype, using the alternate components you suggested. While the resulting device was functional, it failed in
two important aspects. Unit costs would be considerably increased, and the
dimensions would not be in accord with specifications agreed on earlier. I
believe you felt the latter point was less important because you're not aware
of changes we've made in the cabinet design. I have enclosed a copy of
the revised dimensions, and you can see that it is crucially important to
maintain the original specifications. As a result, there just isn't any way we
could adjust our plans to fit the changes you requested. Please proceed in

accordance with the terms of our contract, and if you anticipate further problems in doing so, I hope you will let me know immediately. In that event, I would have to review the entire matter with our chief engineer.

● *RESPONDING TO CLIENT REQUEST FOR PRICE REDUCTION*

General Guidelines

When responding to the type of situation illustrated by Letter 9-2, a basic question needs to be asked: What was lacking in the original agreement that encouraged the client to make his request? Even though the situation calls for tact and diplomacy, the positive side often leads to a strengthening of controls and policies to help avoid future situations of this type. Thus, the client may not be fully placated by your response in a particular instance, but future clients will be less inclined to make similar requests. The reasoning behind your refusal should be explained carefully and diplomatically. If there is a possibility of resolving the request in a more positive way, as in the case of Letter 9-2, indicate this, but do not offer strong assurance unless you feel it is realistic to do so.

Letter 9-2

EPSON FOUNDRY, INC.
820 Worthington Boulevard
Tucson, Arizona 00000

November 7, 19—

Mr. William Mills
Micro Products, Inc.
54 Elton Avenue
Philadelphia, Pennsylvania 00000

Dear Mr. Mills:

Your letter requesting credit for the gear housing we made for you in October has been given to me by our sales engineer.

Although we realize that machinery rebuilding plans are sometimes changed at levels above that of the purchasing agent, we
are sorry we cannot accept the gear housing for credit. It was
cast to your specifications, and, unfortunately, there is no
market for it among our other customers.

If by some remote chance we do get an inquiry for this type of
casting, we will of course get in touch with you immediately. In
the meantime, I'm sorry we can't do more for you.

Sincerely,

Ernest Lipscomb
Engineering Department

Alternate Approaches

(a)   Thank you for sending your letter of June 16 directly to my attention. I have discussed it at considerable length with John Mayer, the design
engineer who worked on this assignment. You may recall during our earlier
discussions that concern was expressed regarding the amount of research
required, prior to completion of the preliminary schematics. The nature of
this research, and the amount of time it required, has been summarized on
the enclosed breakdown of costs. Your inquiry, understandably, was based
only on the data supplied with our bill of June 12. Now, with the additional
and more detailed data enclosed, I hope you agree that the total expense
involved is clearly in line with the original cost estimates. Thank you again
for your letter, Mr. Clayton.

(b)   Your request for a reduction in the fee we submitted to you on May
4 has been received, and I can appreciate your concern in this matter. We
are concerned, too, because we value the relationship our firms have
enjoyed over the years. Your letter contained a clue that enabled us to
pinpoint the cause of this misunderstanding. If you will refer to page 2 of
the price schedule supplied by the outside distributor (see enclosed copy),
you will note the quantity prices quoted were for *5,000 to 10,000* units, in
multiples of *1,000*. The bill you received was based on this price schedule.
It is quite possible that you were not sent a copy earlier, and if that is the
case, I can certainly understand why you would raise a question about
this. Please don't hesitate to send any additional questions directly to my
attention. We want to make certain this matter is resolved as quickly as
possible.

(c)   Your letter of September 18 is appreciated. Not only because it was a courteous and thoughtful request, but also because you are absolutely right. The bill you received was in error. It resulted from one of those infrequent lapses in communications that occur just when you feel all appropriate people have been informed. In this case, it involved increases in our fee schedule for services we provide to industry. This was not an across-the-board increase and certain services were exempted, including the one we provide to your organization. Regrettably, the individual who issued the bill to your firm had not been informed of this. Rest assured the matter has now been corrected, and I have enclosed a revised statement reflecting the cost we agreed on earlier. Again, let me thank you for bringing this to my attention, Mr. Drake.

● *ACKNOWLEDGING CLIENT'S RECOMMENDATIONS*

General Guidelines

On those occasions when a client offers an idea for your consideration, a reply is called for whether you decide to accept or reject the suggestion. As in the case described by Letter 9-3, your response is simplified if the suggestion is a good one and you decide to implement it. In any event, come right to the point when expressing your reaction to the idea, with a tone that reflects sincerity and appreciation. Mention the specific suggestion that was offered, and tell the reader what action, if any, is being taken as a response. If you decide, for whatever reason, that the idea is impractical, explain your reasoning in a way that assures the reader it was given serious consideration. Your closing sentence should once again express your thanks for the writer's thoughtfulness.

Letter 9-3

MICRO ELECTRONICS CORPORATION
416 Royal Street
Waycross, Georgia 00000

March 21, 19—

Ms Doris Henderson
Ames Technical Institute
216 Doe Road
Elberon, Georgia 00000

Dear Ms Henderson:

Thanks so much for your thoughtful letter suggesting that we replace the plain tabs with laminated tabs in our training manual
for the X400 computer.

After checking into it, we agree with you completely. Laminated
tabs do indeed stand up much better under constant usage by our
computer customers. The additional cost is minimal and is further outweighed by the utility aspects of this improvement.

Consequently, you will be glad to know that as a direct result of
your letter, and beginning with the next printing, our training
manual will have laminated tabs to designate the different sec-
tions in the manual. In addition, as an extension of your idea,
they will be slightly larger so as to permit an expanded description of contents in each section. To show our appreciation for
your fine suggestion, we will send you the first "new look"
edition of the training manual with our compliments.

Sincerely,

JOHN BENTON
Technical Manual Department

Alternate Approaches

(a)   Your letter of May 12 reflected the thoughtful consideration you obviously gave our KX12 model, and we sincerely thank you for this. Our technical staff always appreciates feedback from clients on product refinements, particularly when they come from someone with an engineering background. In this case, I want to be completely frank with you about our reaction to the change in dimensions you referred to. The decision not to implement the suggested change is based on a factor you, understandably, are not aware of. At a staff meeting held some weeks ago, it was decided to combine this unit with the R10 chassis and discontinue production of the KX12 as a separate product. However, the reasoning behind your suggestion was sound, and we intend to see if the thought might be applied to other, somewhat similar products we manufacture. The fact remains, however, we very much appreciate the time you took to write us about this, Mr. Holbrook. Once again, our sincere thanks.

(b)   You have not only my thanks for your letter of June 4, but our technical staff joins me in expressing our appreciation. The intriguing suggestion described in your letter offers a unique approach to a potential problem that never occurred to us. As you know, the passive infrared sensor you referred to requires 12 V DC at 20 mA. The adjusted rating you described would require modification of the power source, and this will involve both engineering research and market studies to determine if such a change would be cost effective. Frankly, we're not certain it would be, but you can bet we're going to find out. Rest assured, you will hear from us again as soon as a determination has been made. But in the meantime, I wanted you to know that we received your letter and that we are grateful you took the time to write us about this.

(c)   The idea suggested in your letter of March 6 is well worth serious consideration, and you may be sure that is what it will receive. The lubricating gel you referred to has not been tested by our staff, but we have ordered a supply and will move ahead on appropriate tests as soon as it is received. While I understand your firm manufactures this product, the use you suggested is one that intrigues us because of the assertion that the product does not attract dust. We have been assured the test supply will arrive within the next two weeks, and tests will begin shortly afterward. Our plan is to apply the gel to 100 units and measure the results over a one-month period. If the results are as satisfactory as all of us hope they will be, you may be certain the procedure will be standardized and used on several of our products. In the meantime, thanks very much for writing to us about this, Mr. Carson.

● *RESPONDING TO REQUESTS FOR CHANGES IN CORPORATE POLICY*

General Guidelines

   Operational policies and procedures, no matter how carefully developed, are weakened to the extent exceptions are made. Letter 9-4 deals with one form of request for an exception, and the alternate situations deal with others. The originator of the request may be a staff member, or it may come from a particularly important client. In either event, the response must be clear and unequivocal. On those occasions (rare, it is hoped) when an exception is clearly justified by the facts, it's a good idea to emphasize the exceptional nature of the situation and the fact you believe the basic policy is both fair and reasonable. Regardless of your decision in a particular case, it should be expressed courteously and clearly reflect the reasoning you have applied.

Letter 9-4

WAYCROFT PRODUCTS, INC.
516 Klein Terrace
Wayside, Illinois 00000

April 25, 19—

Mr. Robert Johnson
2032 Kenyon Road
Suburbia, Mississippi 00000

Dear Mr. Johnson:

Comparisons will show that Waycroft Products has one of the best warranties in the field, giving the widest coverage, but we do not cover the use of our Precision Drill No. 419 in any manner not specified in the warranty. If we did so, the price of the drill would have to be increased astronomically.

Concerning the drills you referred to, our technical staff has determined that all the chucks were damaged when the drills were

```
used on materials clearly excluded by the warranty. We can re-
place the chucks at cost to you, including postage, and return
them to you in approximately two weeks. If you would like us to
do this, let me know, and I will have someone on our staff give
you precise cost and time estimates. If you will send your re-
sponse directly to my attention, I will see that the appropriate
action is taken.

Sincerely,
```

Alternate Approaches

(a) Corporate policies are not established arbitrarily. They are designed with the welfare of all our personnel in mind. As a practical matter, they cannot be built around the individual interests of each staff member. The result would be chaos. This is not to say I don't sympathize with your request for a change in our observance of holidays. Looking at it from your standpoint, I can understand why you would make the request. Yet, if we took this action, every other staff member would have a right to expect a similar response to their individual requests. So, Tom, I'm afraid we cannot go along with the policy change you requested. While it may not appear so to you now, the fact we want to be fair to the greatest number of employees will ultimately work to your benefit, as well as to everyone else's.

(b) Thank you for your letter of May 6. I'm genuinely sorry about your experience with our RS232 connectors. Corporate policy in a situation of this type would normally prevent us from accepting return of the cable connectors, but under the circumstances we are making an exception in this case. Despite the fact the return privilege time period has expired, we agree that the literature you received should have been clearer on the price differential between solderless- and solder-type connectors. In a sense, you have done us a favor by bringing this to our attention, and it is only fair that we reciprocate. Please return the shipment directly to my attention. We will credit your firm for the entire amount, including return shipping costs. Thanks again for your letter, Mr. Blakely.

(c) As are most people, I am not especially fond of the phrase "It's against company policy." Yet I suspect most managers at one time or another find it necessary to use either that phrase or it's equivalent. In this instance, I would be less than frank if I didn't tell you our prepayment fee schedule must be applied in this case. The procedure is one that is carefully and fairly applied to all our accounts, and making an exception of the type requested in your letter would be unfair to other firms who retain our services. It is the desire to be fair to all our clients that motivates this response, Mr. Wells. It is not an arbitrary position we take only with recent clients. Please let me know if there are other ways I might be of help to you. Despite the position I must take in this matter, we would sincerely like to be of service to you and your firm.

● *REFUSING TO ACCEPT RESPONSIBILITY FOR AN ERROR*

General Guidelines

Only the masochist enjoys being blamed for a mistake he or she did not make. When disclaiming responsibility for an error, the most basic requirement is to make absolutely certain of your facts. Note two additional elements in Memo 9-5 that are also important: the denial should not be equivocal or open to misinterpretation, and the tone should not be accusatory. To avoid winning the battle and losing the war, a genuine effort should be made to defuse the situation. If shifting the blame to the real culprit becomes essential, do so with as much objectivity and restraint as possible. An emotional or vindictive response does more harm than good.

Memo 9-5

April 25, 19—

TO: Robert Jones, General Manager

FR: Henry Karlin

I am sorry that production has been disrupted in Department M4, but that department's earlier instructions were followed to the letter. Hindsight indicates that they might have been interpreted two ways, and I am enclosing a photocopy of the instructions to show you what I mean. Under the circumstances, I think you can understand why we do not feel it is reasonable for us to accept full responsibility for the error.

Although it has never been required up until this time, I think
in the future I will always double-check important requests like
this by telephoning the department head. Of course this does not
help the present situation, but if you or Mr. Davis can think of
any way I can assist in unraveling the snarl, I shall be most
happy to do so. Please let me know.

Alternate Approaches

(a) We are genuinely sorry that you did not receive the diagrams you requested for your display at the Tech-Mart Conference. The reason is a simple one, even though the outcome led to complications. Please note the copy of your original request enclosed with this letter. As you can see, someone inadvertently wrote down the wrong serial number. When our design supervisor received this, she immediately referred to the diagram listing we sent you on March 12. Since the number written on the order also appeared on that list, she sent the diagrams you received. While the circumstances in no way minimizes our regret over the situation, I did at least want you to know why the error occurred. You may be certain we value our relationship, Mr. Carson, and if you would still like to have the diagrams indicated in your letter of March 22, I'll be glad to rush them to you via Federal Express.

(b) Your letter came as quite a shock, and my initial reaction was to feel as much anger toward our people here as you must have felt. You can bet we moved quickly to trace the origin of this problem. The cause turned out to be something you probably had no way of knowing. Here are the facts: the anodized tubing was definitely returned within the 30-day guarantee period. You are absolutely right about that. But you are not aware, and neither was I at the time your letter arrived, that four $3/16''$ holes had been drilled in each one, possibly for later installation of connecting brackets. It was at that point your staff discovered they needed 16′ lengths, not 14′. Under the circumstances, Mr. Allen, I hope you can understand why the refund was not made. If there is some use you can make of the tubing, I will, of course, be glad to return it to you at our expense.

(c) Even though a thorough investigation shows that Mrs. Harvey, was not at fault, I feel compelled to express our sincere regret over this entire episode. I have reviewed the original request from your firm and compared this with the research data provided to you by Mrs. Harvey. Your staff had originally requested data resulting from studies made in four states: New Jersey, New York, Pennsylvania, and Maryland. Mrs. Harvey tells me that on May 4, a Mr. Walton from your firm called to request that the state of Delaware be substituted for Maryland. Despite the fact she had started to compile data for Maryland, Mrs. Harvey discontinued further efforts in this direction and accelerated her efforts to complete all the research within the time period we agreed on earlier. She was able to do this, and the data were delivered to your firm one day ahead of schedule.

● *REJECTING REQUESTS FOR ADDITIONAL TIME OR MONEY*

General Guidelines

The successful completion of various projects can be adversely affected by inaccurate estimates as to the amount of time or financial support required. In accordance with Murphy's law, this frequently comes to light at the worst possible time, when you are faced with an inflexible deadline. Getting back on track requires a fast response to the inevitable request for more time or money. The particular circumstances involved will dictate the tone of your letter or memo. Letter 9-6 leaves little doubt about the writer's position; it is stated in the first sentence. When you can suggest a realistic alternate approach, do so, but clearly specify who has the ultimate responsibility for successful completion of the assignment.

Letter 9-6

WORCESTER MANUFACTURING CORPORATION
42 Phoebus Road
Arlington, Colorado 00000

April 25, 19—

Mr. George Green
2418 Ames Street
Boulder, Colorado 00000

Dear George:

I regret to tell you that the extended time you say will be required on the Garrison job is absolutely out of the question. If

it becomes imperative for you to get outside help on the job,
then consider subcontracting part of it, under your personal
supervision. If you feel action of this type is essential, I'll
see that appropriate members of our engineering staff give it
serious consideration. As you know, I've been impressed with the
quality of your work and would like to see our relationship
continue.

Let me know in the next day or so what you decide, but it is imperative that we keep to the projected schedule.

Sincerely,

Jack Wells
Production Manager

Alternate Approaches

(a)   Having been faced with comparable dilemmas in the past, I can
sympathize with the problem described in your letter of October 2. Regrettably, I must now tell you what our clients told me on those occasions.
Any further delay in delivery of the parts we ordered on September 21 is
unacceptable. In this case, we would have to shut down an entire section in
our Production Department, and I couldn't begin to tell you how many
problems that would cause. The difficulties would not be limited to this
operation, it would inevitably affect our future business relationship with
your firm too. As I see it, Tom, we have two choices. Either your staff
manages to deliver the parts on schedule, or we'll have to apply the
penalty clause. Please let me know by Tuesday of next week how you plan
to handle this.

(b)   While your recent letter was disturbing in many respects, it is fortunate you brought the problem to my attention during the early stage of our
negotiations. Let me come straight to the point so there will be no misunderstanding between us. If your firm cannot operate within the cost limitations specified in my letter of April 12, then both of us are wasting time by
pursuing the matter. Frankly, I would regret this because I wanted to use
this as a test case in determining future assignments we might direct toward
your firm. As I told you earlier, considerable thought and research went
into our cost estimates. Instead of encountering delays at this point, we
want to move ahead as quickly as possible toward the first phase of the

project. Before we decide to explore this with other firms, I would appreciate some final word from you on this.

(c)   Thank you for your forthright letter of May 2. I will be equally candid in this response. The answer, regrettably, is no. I say regrettably because I had hoped we could establish this first assignment as the basis for a continuing relationship. Your technical staff impressed us during our recent meeting, but I can tell you that management here would never go along with the fee increases suggested in your letter. As you know, we deal with various organizations that provide consulting services. I would much prefer not to open the door to newcomers at this point, but if there is no recourse, then we will take this approach as soon as it becomes necessary. It will neither serve your best interests nor mine to procrastinate. Our chief engineer needs a decision on this no later than the end of this month, and, as a result, I would sincerely appreciate having your response as soon as possible.

● *OFFERING ALTERNATIVES WHEN REFUSING CLIENT REQUESTS*

General Guidelines

In addition to the specific circumstances covered earlier in this chapter, there's a broad range of additional situations that require you to say "no" to a client. For example, Letter 9-7 deals with a rather common request involving delivery schedules. Three fundamental objectives should be considered when drafting this type of letter. First, of course, is to express your negative response clearly and courteously. Next, a brief explanation of your response is required. Finally, reasonable and realistic alternatives should be suggested whenever possible. Bear in mind that the primary purpose of your letter is not only to deny the request, but to retain the client's goodwill while doing so.

Letter 9-7

AINSLEY PRODUCTS, INC.
42 Bay View Avenue
Elton, New Jersey 00000

April 24, 19—

Mr. John Dearing
Component Manufacturers
350 Suburban Road
Industrial Town, Maryland 00000

Dear Mr. Dearing:

I regret that it will be impossible for this department to make deliveries two days in advance of the present schedule. The reason is that the deliveries we make to you depend on components we receive from five different manufacturers. I cannot conceive of our being able to persuade all of them to alter their schedules to accommodate our request. I regret this too and hope the next time you ask a favor I'll be able to grant it.

I do have one suggestion—perhaps not a total solution, but you might want to explore it. If your new setup depends on receiving deliveries on Mondays instead of Wednesdays, would it make any sense simply to save the shipments as they arrive and put them into the pipeline the following Monday? That just might be a possibility you'd want to explore.

With best regards,

Alternate Approaches

(a)   While I agree it would simplify matters if we acceded to your request of March 12, the simplification would be shortlived. Complications would surely follow, and they would far outweigh any initial benefits. As a result, and in your own best interests, we've decided not to provide the additional consulting service you requested. Our expertise is concentrated in that area where we are now serving you, and I'm glad we have the privilege of doing so. To risk an expansion of this into procedures we have had little experience with could endanger our relationship, and that is the last thing in the world I would want to happen. At the same time, we do have a suggestion you may want to consider: Acme Laboratories in Carswell did some work for us in this area last year, and we were pleased with the results. Their brochure is enclosed. Would you like me to ask the representative we dealt with to get in touch with you? If so, I'd be glad to do this.

(b)   Many thanks for your interesting letter of May 6. Your request for a redesign of the hardware we supplied reflects an ingenious use of the brackets involved. So ingenious, as a matter of fact, this is the first time we've heard of someone using them for the purpose you described. Therein lies the problem. While our engineering department would view this as an interesting challenge, the marketing people tell us the demand would be so limited that the retooling involved simply would not be cost effective. Naturally, this view is subject to change, depending on the law of supply and demand. In the meantime, here's an alternate approach you may want to consider. Suppose we reinforced the vertical strut in order to resist pressure in the direction of its length. Would this solve the problem for you? If so, we'll develop the idea in sufficient detail so you'll have cost data, weight limitations, and so on.

(c)   Thank you for returning a representative unit from among the circuits we supplied you with on May 2. As I told you during our telephone conversation, we were greatly disturbed by the news of breakdowns you've encountered with these boards. Now that our technical staff has made a thorough check of all the components, we've finally pinpointed the source of the problem. You'll recall we supplied only partially completed circuit boards, per your request. Your staff added the remaining parts, including the power semiconductors. In essence, the problem is caused by inadequate heat sinks being used for the semiconductors. The TO-220 cases require heat sinks similar to the one I've enclosed. If you would like to return all the boards and have us install the correct heat sinks, we'll be glad to do so at cost. But under the circumstances, I think you can understand why we cannot issue the credit you requested.

● CONVEYING NONCOMMITTAL RESPONSE TO COLLEAGUE'S REQUEST

General Guidelines

Letter 9-8 illustrates the kind of qualified response called for under the circumstances indicated. The situation is such that a definite yes or no answer is not possible, or advisable. This type of verbal tightrope walking requires a delicate touch, if you are to avoid any harmful effect it might have on your relationship with the person you're writing to. Diplomacy is one important element, brevity is another. Long-winded explanations (or excuses) will immediately raise questions as to the sincerity of your response. Your opening sentence or paragraph should set the stage for the position you're about to take. Try to close your remarks with an upbeat or positive reference to your desire to be of assistance if at all possible.

Letter 9-8

SMITH AND ROSS MANUFACTURING
88 Dorian Avenue
Max Meadows, Indiana 00000

April 24, 19—

Mr. Richard Roe
Warner Corporation
912 Ellen Street
Carey, Ohio 00000

Dear Dick:

You have caught me at just the wrong time with your request for help. We are working against several deadlines, and two of our key men are laid up with injuries. I agree that prefabrication might be the answer to some of the cost problems you are running into on the Dominion Estates project. This is not something you can just go into cold turkey, however, and we should discuss it before you make a move in that direction.

How much time do you have on this? I won't be able to make a move in your direction for at least three weeks. Let me know if that will be okay, and I will definitely put it on my calendar. It bothers me that I cannot get together with you immediately.

With best regards,

Alternate Approaches

(a)   It was coincidental that you decided to write me about the problems being encountered with your changeover to automated controls. We have been having a few problems here that are somewhat similar to those you described, and I had intended to inquire about your experience in this area. I think you know I'd like to help in any way I can, but quite frankly, until we come up with solutions that are acceptable to us, I'd be reluctant to offer any untried and unproven suggestions to you. Doing so at this stage could quite possibly do more harm than good, and my desire would be to help, not harm, your current efforts. During a recent staff meeting, it was estimated that within the next few weeks we should complete our testing of revised procedures, designed to correct the problems you described. I should then be able to give you a more definitive, and reliable, response to your questions.

(b)   I was impressed with the detailed nature of your relocation plans, and my initial reaction was to draw general responses from our staff to the questions you asked. Cooler heads prevailed, however, and I believe this may ultimately work in your favor. In essence, your detailed questions deserve answers that are equally specific. I'm not sure we can, but we'd like to give you the kind of in-depth responses that will be of greatest help to you. That, however, will take time. The silver lining to this cloud is that you are still, as you indicated, in the early stages of your planning. That is fortunate for both of us. My guess is that we can complete the necessary research in about two weeks. If we can do it sooner, you can bet we will. In any event, we'll follow through on this as soon as possible, and you'll hear from me as soon as the data have been compiled.

(c)   Your promotion to the senior engineering staff came as great news and you have my heartiest congratulations. It was good of you to attribute some of your progress to the training you received here, Tom, and I appreciate your thoughtfulness. With reference to the questions you asked, I have to give you a somewhat qualified response, because my experience doesn't extend into this area. But I have some contacts that might be able to help. Problem is, they are not on my staff, and it may take a little time to get in touch with them. Rather than have you try to track them down, it seems better if I make the effort and, if they can provide the answers you need, pass this information along to you. It shouldn't take more than a few days, but regardless of whether they can or cannot supply the data you need, I'll get back to you on this as soon as possible.

● *ACCEPTING OUTSIDE BID FOR WORK PROJECT*

General Guidelines

In addition to the legal aspects pertaining to a contractual agreement, the covering letter should be crystal clear and reflect the essence of the contract itself. As in the case of Letter 9-9, if there are particularly important terms covered in the formal agreement, they may also be stressed in the letter. Assuming a properly drawn contract covers all the essential details, the covering letter can be brief and used principally as the transmitting vehicle. If any important points are covered that are not mentioned in the contract, it is suggested that legal counsel be asked to review the letter before it is sent.

Letter 9-9

LARUSO SOLID STATE CONTROLS, INC.
44 Montgomery Street
Elkhart, New Jersey 00000

February 26, 19—

Mr. Robert Charles
Ardsley Services, Inc.
816 Mead Street
Bayonne, New York 00000

Dear Mr. Charles:

The decision has been made to award you the X1400 circuit board project, based on your proposal of January 16, 19—.

We would appreciate your starting immediately to complete the 600 boards in question, consistent with procedures stated in your proposal and in the guidelines that accompany this letter. Let me emphasize again that delivery date of the entire shipment must not be later than May 15. The penalty clause will go into effect if complete delivery is not made on the date specified.

Please sign all copies of the enclosed contract and send them
directly to my attention. I'll return a countersigned copy for
your files.

Sincerely,

Alternate Approaches

(a)   We are pleased to enclose our Purchase Order No. 10422 for the
components listed in your estimate of May 4. In terms of delivery, the
purchase order simply indicates the date we agreed on in our discussion of
May 2. However, I want to add a special note of urgency regarding that
date. Since our meeting on May 2, we have received calls from several
clients alerting us to the possible cancelation of their orders if we do not
complete installation of all equipment by June 16. To do this, delivery of
the components no later than May 2 is imperative. This will still require our
Production Department to go on an overtime schedule in order to complete
the assemblies on time. There is only one way in which we might avoid this.
And that is, if you are able to make your delivery sooner than originally
indicated. Your efforts toward this end would be sincerely appreciated, but
as stated earlier, in no event should delivery be made later than May 2.

(b)   We have made a comparison between prices listed in your letter of
April 12 and those estimated by other suppliers. Although your firm was
not the lowest bidder, the price spread was not that great. The primary
factor that causes us to accept your quotation is your willingness to cut the
metal strips to our specifications without additional charge. This is informal notification that we would like you to move ahead on developing a
final proposal, indicating the earliest possible delivery date. Each item
listed should carry a descriptive reference and your stock number for that
item. Once we have this, and assuming it is in accord with our discussion
and your letter of April 12, I will see that the purchase order is issued.

(c)   I am glad to tell you that your detailed and well-organized proposal
has been accepted. Please carefully review the enclosed contract providing
for 24-hour security guards for our R&D plant in Houston, Texas. The
total fee stated in this two-year contract shall constitute complete consideration for the services rendered by the contractor, except as otherwise
specifically stated in the contract. While the following request is not referred to in the contract, I hope you will be agreeable to it. Specifically,
our research staff should be given detailed information on this security
service, and I believe you are the best one to do this. Or you may prefer to

select someone from your staff to meet with our people. In either event, they should be given a detailed description of the services you will provide. We will set up a meeting for this purpose at your convenience. Please sign all copies of the contract and return them to me. I'll send you a countersigned copy for your files.

● *REJECTING OR ACCEPTING TREATISE ON TECHNICAL SUBJECT*

General Guidelines

While many options are available, one way to reject a technical paper is to explain your reasoning before stating the rejection. Regardless of the approach you use, the response should be courteous and thoughtfully expressed. Quite often, as illustrated in Memo 9-10, a modification of some kind, or additional data on a particular aspect, will raise the quality of the paper to an acceptable level. Obviously, the task is simplified when you decide to accept the paper. Even then, constructive suggestions should not be withheld. They will also assure the author of your support in making his or her presentation even more effective.

Memo 9-10

June 12, 19—

TO: Dr. Richard Nelson

FR: Thomas Riley

Attached are our reviews of your proposed paper on data processing in medicine. I share them with you for the benefit of this paper's development. Based on the reviews, you have the foundation for a fine presentation here. While the concept is sound, it is our opinion that it does not go far enough in the practical applications of the procedures you describe. This limits the value and potential interest in such a work.

Should you be willing to expand the coverage so as to increase its utility value, I would be pleased to reconsider the paper for publication and distribution to our membership. I would also appreciate receiving more information on how your work compares with other research studies in this subject area.

Alternate Approaches

(a) Thank you very much for letting us review your interesting paper on digital radiology. We found it both informative and well organized, particularly in terms of the potential audience you described in the opening section. There appears to be only one area that might be expanded in order to increase the paper's utility value, and I would appreciate your thoughts on this. Specifically, procedures for setting up a picture archiving and communications system, along with practical guidance on the selection of appropriate hardware and software. If this additional section were to match your excellent treatment of compression techniques, for example, clipping and bit truncation, run-length coding, and Huffman coding, we believe the paper would be a fine addition to the literature.

(b) We found your paper on writing technical proposals a bit too specialized for our current publishing program. If it were not oriented almost exclusively toward securing government contracts, our membership would probably express greater interest in your treatment of the subject. Please understand that this response is the reaction of only one association. Opinions vary widely among different groups, and I would urge you to submit the paper to other engineering associations. If you have not done so already, you might also review the *Directory of Technical Publishers.* I have enclosed some literature about this organization that may be of help to you. We do wish you success in locating an appropriate publisher for this work, and thanks once again for sending it to us, Mr. Whitmore.

(c) The editorial board for our association has carefully reviewed your excellent paper on avionics, and I am delighted to tell you it has been accepted for publication. We have seldom seen clearer or more useful guidelines for matching the system to the aircraft, or a better treatment of digital applications in avionics. Your approach to designing for more efficient maintenance procedures was also exceptional. Now, all we need is a more detailed review of your experience in this field. Portions of this data will be included, along with your paper, in the November issue of our *Journal.* Please complete the brief questionnaire enclosed, sign all copies of the publication agreement, and return these papers to me. I'll return a countersigned copy of the agreement for your files. Thank you very much for sending this fine paper to us, Mr. Cary.

● *ACKNOWLEDGING INVITATIONS TO ATTEND CONFERENCES*

General Guidelines

Whether you decline or accept a colleague's personal invitation to attend a professional conference, a brief response is usually all that's necessary. If you are interested, but there are circumstances relevant to the invitation that are not clear, for example, a possible obligation you must assume in order to attend, then your response should be designed to secure the necessary information. Letter 9-11 provides a brief acceptance letter, and the alternate approaches suggest variations you may want to consider. In all cases, particularly in those situations where there is a deadline mentioned, your response should be sent as promptly as possible.

Letter 9-11

RICHARDSON AMPLIFIER CORPORATION
48 Wellsley Boulevard
Hartford, Connecticut 00000

October 17, 19—

Mr. Ross Branch
Abernathy Electronic Devices
Century Park
Brooks, Massachusetts 00000

Dear Ross:

It was thoughtful of you to invite me to the computer fair to be held on December 4, and I accept with pleasure. The agenda for the evening sounds interesting, and I'm looking forward to seeing you again.

```
I'll plan to get there a little before 5:30, and perhaps we can
have a brief chat before the program begins. I don't want to miss
any of the demonstrations put on by the hardware and software
vendors.

Again, thanks very much for the invitation.

Sincerely,


Gerry
```

Alternate Approaches

(a)   Your invitation to attend the symposium on March 12 is appreciated. I assume this will involve my attendance as an interested "listener"
and not as an active participant in the conference discussion. Provided this
assumption is correct, I would be delighted to attend. My purpose in making this clear is to avoid having you assume I am more experienced in this
field than is actually the case. I view it as an opportunity to become better
acquainted with recent developments in the field, and for this reason I'm
grateful that you thought of me. Perhaps as I become more familiar with the
subject area I'll be able to attend future sessions as a more active participant. Thanks again, Tom. I look forward to seeing you and your colleagues
on March 12.

(b)   I regret that circumstances will probably prevent my attending the
conference on May 4 being sponsored by your association. Actually, I had
fully intended to be there, but one of our clients on the West Coast has a
serious problem that must be resolved quickly. This requires me to leave for
Los Angeles on May 1, and I anticipate being there for five or six days. In
the event my presence is required for only a day or two, I'll try to make it
back in time for the meeting. In the meantime, I at least wanted to let you
know that I appreciate your thoughtful invitation.

(c)   I have three responses to your kind invitation to attend the regional
association seminar on March 22. First, of course, my sincere thanks for
inviting me, even though I am not a member. Second, we've recently added
four members to our staff and two of these engineers have had long and
varied experience in the field of fiber optics. I mentioned this meeting to
them, and they are most anxious to attend. Would it be possible to bring
them along with me? Finally, and perhaps I'm really pushing my luck now,

there is the possibility that a transcript of the meeting might be condensed and published in our quarterly *Journal*. Would the association be amenable to this? Naturally, the final text would be subject to editing and approval by your committee, and we would give full credit to your organization. Since time is growing short, I'll call you on Monday of next week so we can discuss this further. Thanks again for the invitation, Bill. I look forward to seeing you again on March 22.

● *HIRING OR REJECTING APPLICANT FOR STAFF POSITION*

General Guidelines

A written notification that confirms the offer of employment is appreciated by the applicant. It is also an opportunity to eliminate any doubt or confusion regarding such details as to where and when to report for work. In the type of reverse situation reflected by Letter 9-12, the decision not to employ a particular applicant can be softened somewhat by a thoughtful explanation of the reasoning behind the decision. There are practical reasons for such a letter, especially with applicants who made a favorable impression and possess above-average credentials. An occasion may develop in the future when you would like to reopen discussions with the individual. If that should happen, your letter would contribute to a more favorable climate at that time.

Letter 9-12

NORTH SLOPE MANUFACTURING CORPORATION
P.O. Box 169
Barrow, Alaska 99723

February 24, 19—

Mr. John Wainwright
160 East 10th Street
Lennon, California 00000

Dear Mr. Wainwright:

After a great deal of deliberation we have decided to select
John Burney from the San Francisco plant as the new chief design

engineer for this division. It was an extremely difficult decision because of the number of experienced people who applied. I'm
certain that with your determination and skill level, the future
will continue to be a bright one for you. The principal reason
for the decision in this case was the fact John Burney's training
in solid-state controls was considerably greater than that offered by other candidates.

I do want to thank you for the time you took to discuss the position, and for your interest in our division. Provided other
opportunities develop that require someone with your qualifications, you can be certain I'll get in touch with you. In the
meantime, my best wishes for your continued progress toward the
goals you have set for yourself, Mr. Wainwright.

Sincerely,

Alternate Approaches

(a)   It is with genuine pleasure that I confirm our telephone conversation of this morning. With the unanimous agreement of all those who met
with you last week, we have decided to offer you the position of systems
engineer with our Farmington plant. As we discussed in considerable detail, the challenges are great in this particular position. But, being a firm
believer in the adage that every problem is an opportunity, the future can
indeed be a bright one for the right individual—and obviously, we believe
you are the right individual. You will discover that every person on our
engineering staff is an important member of a team that works together.
Continuing success for the staff as a whole depends on the individual success of each member. We believe you will contribute to this success, and we
look forward to having you with us. Congratulations, John, and I'd appreciate it if you'd stop by my office first when you report for work on June 1.

(b)   I would like to express our sincere appreciation for your courtesy
and cooperation during the interviews conducted last week. Quite frankly,
we were impressed with your qualifications. You probably guessed, as a
result of some of the discussions, that the only point of major concern had
to do with your limited experience in our field. This became a determining
factor, and I sincerely regret we cannot offer you a position at the present
time. Under the circumstances, we will continue to search for someone
with experience more closely allied to our operation, but we intend to keep
your application on file for the next six months. It is my hope that a more
suitable vacancy will develop during that time. Provided that happens, you

may be certain we will get in touch with you. If, at that time, you are still interested in joining our firm, I would be glad to meet with you again.

(c)   Congratulations on being selected as our new quality control director, starting on May 12. I hope you will be as pleased as we are with this decision and that future events will measure up to both our expectations. The choice was difficult because so many fine applicants were considered. However, your extensive experience and the enthusiasm you displayed won out over the others.

Personally, I feel the right choice was made, and feel certain the other managers here agree. A copy of the employee agreement is enclosed, and we'll discuss this and other details when you report for work. Relocation expenses will be reimbursed in accordance with the arrangement described earlier, and tentatively let's plan on your arrival here no later than March 15. In the meantime, the *Staff Manual* will be sent to you separately, and I'd appreciate it if you go over this carefully. If questions develop regarding the manual, or any other aspect of our discussions, just give me a call. We look forward to having you with us, Bill.

A WORD ON STYLE

"Please Do Not Hesitate To . . ."

Some books on letter writing have long lists of overworked phrases to avoid. The only trouble is that most of us have long since decided to avoid using such archaic phrases as "Responding to yours of the 18th instant" or ending them "I remain, therefore, your obedient servant."

Nevertheless, there are a few phrases of more recent vintage that tend to be overworked. One of them is "Please do not hesitate to write (phone, call, contact me, let me know, or whatever)." Actually, this is a very comfortable phrase. It serves its purpose. It neatly straddles the line between being formal and being helpful. And nobody seems to object strongly to it. It is seldom encountered in usage books of no-no's.

A number of letters in this book use the phrase. The reason is that many excellent letter writers include it in their last or next-to-last sentences. Because of that, alas, the phrase is fast becoming a cliche. If you want your letters to sound somewhat more original and personal, you may hesitate to end them with "Please do not hesitate to . . ." There are alternatives, as other letters in this book illustrate.

Chapter 10

MASTER CHECKLIST OF ESSENTIALS FOR THE SUCCESSFUL LETTER

Guidelines to Stationery Selection: Judging the quality of paper . . . How to order stationery . . . Envelopes for letterheads . . . Continuation sheets . . . Letterheads for air mail. . . . The Structural Parts of a Letter: The date line . . . The reference line . . . The inside address . . . The attention line . . . The salutation . . . The subject line . . . The body of the letter . . . The complimentary close . . . The signature . . . Identification initials. . . . Letter Styles: The format . . . Complimentary close . . . First sentence and opening paragraph . . . Format . . . The last sentence . . . Letterhead . . . Paper . . . Paragraphs . . . Salutation . . . Signature guidelines . . . Style . . . Tone

The impact of a well-written, attractively "packaged" letter is well worth the effort to produce it. Conversely, a skillfully written letter may not do its job properly if you ignore its physical appearance. A letter on a poorly designed letterhead, reproduced on cheap stationery, and typed in an illegible font will almost certainly make an unfavorable impression.

The design of the letterhead, the quality of the stationery, the style and neatness of the typing are all extremely important. It is often the first impression the reader gets of your letter that determines the response.

This chapter presents easy-to-follow and precise guidelines, all designed to help give your letter the type of efficient, well-organized appearance that will reinforce your message. It also provides helpful hints on structuring your letters and memos so that the end product communicates your message clearly and concisely.

● *GUIDELINES TO STATIONERY SELECTION*

Here are four key factors to consider when selecting stationery:

1. The *feel* of the paper is one of the standards by which letterhead papers are evaluated. High-grade papers have good bulk, crispness, and crackle like a new bank note.
2. The *workability* feature of the paper includes the ease of use, economy of operation, and quality of workmanship the paper provides.
3. The *permanence* feature of the paper is judged by its ability to reproduce clearly defined characters, to withstand erasures, and to permit clear, smooth-flowing signatures.
4. The *cost* feature of the paper is an important factor since it goes into the overhead of the particular office. Management must decide whether the cost justifies the type of letterhead it wants to use for business correspondence.

Judging the Quality of Paper

Weight. Paper weight is measured by the ream. Each ream consists of 500 sheets, 17″ by 22″. There are 2,000 business letterheads of the 8½″ by 11″ size in a ream. For example, "substance 16" letterhead means 2,000 sheets, letterhead size, weight 16 pounds. The weight of a ream may be 12 pounds, 24 pounds, or even more. The better the quality of paper, the more it weighs.

The 20-pound weight paper is the most commonly used paper for business letters; the 16-pound paper is widely used as general-purpose paper within the company. Some executives' letterheads are printed on very fine paper of 24-pound weight.

Content. Paper used for business purposes is made from wood pulp, from cotton fiber, or from a combination of the two. Letterhead paper made from wood pulp is called sulphite bond. It is manufactured in grades 1, 2, 3, and 4, with grade 1 being the best. The most durable, the best looking, and the most expensive paper is that with a 100 percent cotton content, that is, made entirely from cotton fiber. Formerly, the watermark "rag content" was used by all fine paper mills to describe their better grades of paper containing cotton fibers. The watermark "cotton content" is being used today to describe more adequately the basic fiber contained in fine papers.

Grain.   All paper has a "grain," or chief fiber direction. It comes from the process by which the paper is made. In letterhead paper the grain should be parallel to the direction of the writing. (Letterheads to be used with one of the duplicating processes are an exception.) The sheets hug the typewriter platen better and provide a smoother, firmer surface for the impression. Erasing is also easier with the grain parallel to the platen. Every sheet of paper has a "felt" side and a "wire" side. The letterhead should be printed on the felt side.

These criteria for quality paper apply whether your secretary is using a typewriter or a word processor. In other words, even if your secretary is typing your letters into a computer-driven word processor, the letterheads that are put into the printer to print out (via the computer) the letters you dictated should be on quality paper in order to present a good impression of your company.

How to Order Stationery

Even if you get all the specifics over the telephone, ask for a written quotation on cost and on the weight and content of paper, as well as the kind of engraving or printing that will be on it.

In placing the order, specify

1. the quality in reams or sheets
2. the weight
3. the content
4. the grain
5. that the letterhead shall be printed on the "felt" side of the paper
6. the size
7. the color
8. the previous order number, if it's a repeat order
9. the supplier's reference number, if it's an initial order
10. that the letterhead shall be identical (or similar) with the attached sample, if that's what you want

Envelopes for Letterheads

Envelopes of the same quality as the letterhead paper add to the good impression of the company. However, with today's high cost of quality envelopes, more and more companies choose lesser quality paper for envelopes which are only opened and thrown away by the secretaries. Generally, it's only the letterhead that reaches the addressed executive.

The most popular envelope sizes for businesses are

No. 5	4 1/8″ by 5 1/4″
No. 6 1/4	3 1/2″ by 6″
No. 6 3/4	3 5/8″ by 6 1/2″
No. 10	4 1/8″ by 9 1/2″

No. 6 3/4″ and No. 10 are suitable for 8 1/2″ by 11″ letterheads. The Monarch size, 3 7/8″ by 7 1/2″, is used with the smaller executive stationery. No. 5, which is the Baronial size, is also widely used by business executives for their personal stationery.

Continuation Sheets

These sheets, used for second and subsequent pages, should be of the same size and quality as the letterheads. Continuation sheets and letterheads should be ordered together. Comparatively few letters run to more than one page, but the percentage varies with the office. After a secretary has placed two or three orders for stationery, he or she will be able to judge fairly accurately the proportion of continuation sheets to letterheads.

Letterheads for Air Mail

Many companies specify onion-skin paper for lightweight business letterheads used for foreign air mail correspondence. However, letterheads on lightweight paper made from clean, strong cotton fibers are attractive and can withstand frequency of handling.

● *THE STRUCTURAL PARTS OF A LETTER*

Although today's business letters are rather informal, there are certain conventions your secretary should not ignore. Modern practice still requires the use of certain essential components in business correspondence, as well as one or more optional parts.

essential	*optional*
date line	reference line
inside address	attention line
salutation	subject line
body	identification initials
complimentary close	
signature	

The following section takes each element of a letter, in the order of its position on the page, and explains its proper usage and form.

The Date Line

Every business letter you send out must be dated. An efficient secretary will take care of this automatically, without you dictating the date line. Your only job is to check—when signing the letter—that she is doing it correctly.

Where the date line goes.   The standard position of the date line on the page is two to four spaces below the last line of the letterhead, flush with the right-hand margin of the letter. However, the full-block and the simplified letter forms should have the date flush with the left-hand margin. If the letter is very short, the date line may be dropped to give a better balance to the page.

With some letterheads, the letter presents a better appearance if the date line is not placed in the standard position. Sometimes the date line is centered on the page, two spaces below the letterhead. This style should not be used unless the date line is easily distinguished from the letterhead.

How the date line is typed.   The typed date line should be in accordance with the following standard procedures:

1. The letter should be dated the day it is dictated, not the day it is typed.
2. The complete date goes on one line.
3. Do not use *nd, st,* or *th* following the day of the month.
4. Do not abbreviate or use figures for the month.
5. Do not spell out the day of the month or the year, except in very formal letters.

right	*wrong*
September 10, 19—	September 10th, 19—
	9/10/—
	September fifteenth,
	Nineteen hundred and eighty_____

The Reference Line

If a file reference is given in an incoming letter, include a reference line in your reply. Your reference goes beneath the incoming reference. If your letterhead includes a printed reference notation, such as *In reply please refer to,* your reference line comes after it. Otherwise, the reference line comes four spaces beneath the date.

Your Reference #2032
Our Reference #4796

The Inside Address

The name, the addressee's title, the name of the company, and the address make up the inside address.

Where to type the inside address. The proper placing of the inside address is important because its position makes an impact on the format of the letter. If the inside address is set too far to the left or right, a corresponding placement of the right-hand margin may make the letter too squat or too thin, If not placed high enough, the inside address may destroy the vertical spacing of the letter and make a second page necessary.

Whether you are using the traditional or indented style, or today's most popular format, the block style, the inside address should not begin less than 2 spaces, nor more than 12 spaces below the date line, depending upon the length of your letter. The first line of the address should not extend beyond the middle of the page, however.

How the inside address should be typed.   There are several rules your secretary should follow in typing inside addresses.

1. The inside address should correspond exactly to the official name of the company you're writing to. If *Company, Co., The,* or *Inc.* is part of the company's official name, use the form that is in the company title.

2. Do not precede the street number with a word or sign.

right	*wrong*
70 Fifth Avenue	No. 70 Fifth Avenue
70 Fifth Avenue, Room 305	#70 Fifth Avenue
	Room 305, 70 Fifth Avenue

3. There is no set rule about abbreviating words that stand for street direction, such as S. for South and W. for West. But it is a good idea to abbreviate only when there is a good reason, such as to shorten a long line in the inside address.

4. Spell out the numerical names of streets and avenues if they are numbered 12 or under. When figures are used, do not follow with *d, st,* or *th.* Use figures for all house numbers except *One.* Separate the house number from a numerical name of a thoroughfare with a space, a hyphen, and a space. Authorities give different rules for writing addresses, but the following are standard, approved forms.

23 East Twelfth Street	2 Fifth Avenue
23 East 13 Street	234-72 Street
One Park Avenue	

5. Never abbreviate the name of a city. States, territories, and possessions of the United States may be abbreviated. The following are the approved forms of abbreviations for states, territories, and possessions:

Alabama ......AL	Kentucky........KY	Ohio..........OH
Alaska ....AK	Louisiana ....LA	Oklahoma .....OK
Arizona ......AZ	Maine...........ME	Oregon........OR
Arkansas ......AR	Maryland........MD	Pennsylvania....PA
California .....CA	Massachusetts ...MA	Puerto Rico ....PR
Colorado ......CO	Michigan.........MI	Rhode Island....RI
Connecticut....CT	Minnesota.......MN	South Carolina..SC
Delaware ......DE	Mississippi ......MS	South Dakota ...SD
D.C. ....DC	Missouri ....MO	Tennessee .....TN
Florida........FL	Montana ....MT	Texas ....TX
Georgia ......GA	Nebraska ....NB	Utah..........UT
Guam........GU	Nevada..........NV	Vermont.......VT
Hawaii........HI	New Hampshire ..NH	Virginia........VA
Idaho ....ID	New Jersey ....NJ	Virgin Islands...VI
Illinois IL	New Mexico .....NM	Washington....WA
Indiana........IN	New York........NY	West Virginia ..WV
Iowa ....IA	North Carolina...NC	Wisconsin ......WI
Kansas........KS	North Dakota....ND	Wyoming......WY

Do not abbreviate the following: Samoa

6. The ZIP Code is the last mandatory item in the address. It follows the city and state address.

7. If there is no street address, put the city and state on separate lines.

8. Do not abbreviate business titles or positions, such as president, secretary, and engineering director. Mr., Ms, Mrs., or Miss precedes the individual's name even when the business title is used.
   If a person's business title is short, place it on the first line; if the title is long, coupled with a long name, place the title on the second line.

   Mr. James F. Lambert, Chief Engineer
   Lambert & Woolf Company

   Ms Joan F. Littleton
   Technical Assistant
   Paton & Gibson, Inc.

9. In addressing an individual in a company, corporation, or group, place the individual's name on the first line and the firm's name on the second line.

10. Do not hyphenate a title unless it represents a combination of two offices, such as *secretary-treasurer.*

11. When writing to the officer of a corporation who holds several offices, use only the title of the highest office, unless his or her letterhead states it differently.

    Note that the president of a firm normally is considered subordinate to the chairman of the board of directors. All other officers are ranked in their relation to the president. In cases where an executive is both an officer and a member of the board of directors, correspondence should be addressed to him or her in the officer's capacity, unless he or she has indicated otherwise.

12. If a letter is addressed to a particular department in a company, place the name of the company on the first line and the name of the department on the second line.

The Attention Line

Business letters may be addressed to the company rather than to an individual. However, if you wish to have the letter directed to a particular individual in the firm, you may use an "attention line." The advantage of using such a device is that the letter is immediately recognized as a business rather than a personal one and will be opened in the absence of the person to whom it is directed.

Position of the attention line.   The attention line should be typed two spaces below the address. The attention line has no punctuation and is not underscored. The attention line may appear in either of the following positions:

Thompson Industries
1800 Plandome Road
Detroit, Michigan

Attention: Mr. Joseph P. Roberts

        or

Thompson Industries
1800 Plandome Road
Detroit, Michigan

                              Attention: Ms Alice J. Varga

It is permissible to direct the letter to the attention of an individual without including his or her given name or initials if you don't know them.

(Of course, you can always call up the company and find out the individual's full name from the switchboard operator.)

preferable

Attention: Mr. George H. Richards

permissible

Attention: Mr. Richards

The Salutation

The salutation should be placed two spaces below the inside address, flush with the left-hand margin. If an attention line is used, the salutation should be placed two spaces below the attention line.

The following is the correct way your secretary should type the salutation:

1. Capitalize the first word, the title, and the name.
2. Use a colon following the salutation. A comma is used only in social letters.
3. Mr., Mrs., and Dr. are the only titles that are abbreviated.

Forms of salutation.   The form of salutation varies with the tone of the letter and the relationship between the writer and the addressee. See the chart on page 000 for salutations of varying degrees of formality, together with the appropriate complimentary close to use with each.

You may use a personal salutation or complimentary close in your correspondence to close friends. Such salutations as Dear Jim or Dear Jennifer are appropriate when you know the recipient very well.

Here are the rules that will help you dictate salutations correctly:

1. If the letter is addressed to a man, make the salutation singular, for example, *Dear Mr. Ross* or *Dear Sir.* If the letter is addressed to a company or group of men, make it plural, for example, *Gentlemen* or *Dear Sirs.*
2. If the letter is addressed to a woman, make the salutation singular, for example, *Dear Ms Ross* or *Dear Madam.* If the letter is addressed to a company or group of women, make it plural, for example, *Ladies* or *Mesdames.*

3. Never use a designation of any kind after salutation.

right	*wrong*
Dear Mr. Condon:	Dear Mr. Condon, C.P.A.:

4. Never use a business title or designation of position in a salutation.

right	*wrong*
Dear Ms Broder:	Dear Secretary:
	Dear Secretary Broder:

5. Follow a title with the surname:

right	*wrong*
Dear Professor Stanton:	Dear Professor:

6. The proper salutation in a letter that is not addressed to any particular person or firm, such as a general letter of recommendation, is *To Whom It May Concern.* Note that each word begins with a capital.

7. The salutation in a letter addressed to an organization composed of men and women is *Ladies and Gentlemen;* to a man and woman, *Dear Sir and Madam;* to a married couple, *Dear Mr. (or Dr.) and Mrs. Marsh.*

The Subject Line

There is a growing trend in many companies to use a subject line following the salutation. When a subject line is used, it becomes unnecessary to use a lengthy opening sentence to explain the purpose of the letter. Further, the subject line assists the reader in his or her reference to previous correspondence on the same subject and also acts as an aid to efficient filing.

How to type the subject line. The subject line should be centered two spaces below the salutation. However, when you use the full-block or simplified letter style, the subject line should be flush with the left-hand margin.

Never place the subject line before the salutation. It is part of the body of the letter, not of the heading.

Generally, the subject line is preceded by *Subject* or *Re* followed by a colon. When the word *Subject* is omitted, the entry should be underscored. If there is more than one line in the subject, underscore the last line only. Be sure to capitalize all the important words in a subject line. The following two examples are the preferred methods for typing subject lines:

June 13, 19—

Martin, Hill, Inc.
Rochester, New York

Gentlemen:

Subject: Child Care Booklet

or

June 13, 19—

Martin, Hill, Inc.
Rochester, New York

Gentlemen:

Re: Child Care Booklet

The Body of the Letter

Your secretary will avoid making errors in style by using the following rules as his or her guide:
How to type the body.

1. Single space unless the letter is very short.
2. Double space between paragraphs.
3. When the block style is used, begin each line flush with the left-hand margin of the letter.
4. When the indented or semiblock style is used, indent the first line of each paragraph five to ten spaces.
5. Always indent paragraphs when a letter is double-spaced.

How to write dates within the letter.   When the day precedes the month, it is permissible to write the day out or to use figure. For example, fifth of March, or 5th of March. If the month precedes the day, the figure is always used. For example, March 5.

The importance of consistency in spelling.   Some frequently used words in business correspondence, especially as it relates to computer data processing, have two accepted spelling forms, such as *in-house data processing, inhouse DP; stand-alone systems, standalone systems.* If your firm has not adopted a particular standard, the choice is yours. But be consistent! A letter stating in the first paragraph that the company has decided to do its data processing *in-house,* and two paragraphs later justifying the decision by saying that according to a feasibility study the *inhouse* DP will be more economical, will not leave the reader with a favorable impression.

The Complimentary Close

Where to type the complimentary close.   The complimentary close should be typed two spaces below the last line of the letter. It should begin slightly to the right of the center of the page, except in the full-block and simplified letter. The complimentary close should never extend beyond the right margin of the letter. In letters of more than one page, at least two lines should be on the page with the close.

How to type the complimentary close.

1. Capitalize only the first word.
2. Follow the complimentary close with a comma. This is a better practice even when open punctuation is used in the inside address.

Forms of complimentary close.   The form varies with the tone of the letter and the degree of familiarity between the writer and the addressee. The degree of formality of the complimentary close should correspond with the salutation. The chart on page 332 gives the appropriate complimentary closes for various salutations.

Do not confuse *respectfully* with *respectively.* The latter means with respect to each of two or more, in the order named.

The Signature

What to include in the signature.   Your signature on a business letter usually consists of your written signature, your typed name and business title, and the name of your division or department, if appropriate. Your typed name can be omitted if it appears on the letterhead.

The inclusion in the signature of your business title or position indicates that you are writing the letter in your official capacity. Thus, if you write a letter on the firm's stationery about a purely personal matter, your position is not included in the signature.

VARIOUS SALUTATIONS AND APPROPRIATE
COMPLIMENTARY CLOSES

	Salutations	*Complimenary Closes*
Very Formal	My dear Sir:	Respectfully,
	Sir:	Yours respectfully,
	My dear Madam:	Respectfully,
	Madam:	Yours respectfully,
Formal	Dear Sir:	Very truly yours,
	Dear Madam:	Yours very truly,
	Gentlemen:	Yours truly,
	Ladies:	Very truly yours,
Less Formal	Dear Mr. Suchard:	Sincerely,
	Dear Ms Sheean:	Sincerely yours,
	Dear Dr. Lubow:	Yours sincerely,

(Most business correspondences use these forms rather than the formal forms.)

Personal	Dear Mr. Suchard:	Yours cordially,
	Dear Ms Sheean:	Cordially,
	Dear Dr. Lubow:	Cordially yours,

(Implying personal acquaintance or previous friendly correspondence.)

Engineering firms, attorneys, certified public accountants, and the like frequently sign letters manually with the firm's name, particularly if the letter expresses a professional opinion or gives professional advice.

Where to place the signature. The firm's name should be typed two spaces below the complimentary close, the writer's name four spaces below the firm's name, and the writer's position either on the same line or on the next line. When the firm name is not included, the writer's name and position should go four spaces below the complimentary close.

When the inside address is typed in block form, the signature should be aligned with the first letter of the complimentary close. When the indented form is used in the inside address, the signature is aligned with the third or fourth letters of the complimentary close. In either case, the lines of the signature should be blocked, unless an unusually long line makes this arrangement impractical.

No line of the signature should extend beyond the right-hand margin of the letter.

How the signature should be typed.

1. The signature should appear exactly as you sign your name.

right	*wrong*
Helen P. Dunnel	*Helen P. Dunnel*
Helen P. Dunnel President	H. P. Dunnel President

2. Business titles and degree letters follow the typed signature. No title precedes either the written or typed signature.

The secretary's signature.   When your secretary signs your name to a letter, his or her initials should be placed immediately below it.

Yours truly,

Lyle R. Deen, M.B.

Lyle R. Deen, M.B.

When the secretary signs a letter in his or her own name, your name, official title, and division or department should be typed below his or hers.

right	*wrong*
Secretary to Charles P. Doyle Director Electronics Division	Secretary to C. P. Doyle Electronics Division

Identification Initials

The identification line shows who dictated the letter and who typed it. The only purpose of the identification line is for reference by the business organization writing the letter.

If your company requires the usual identification line, the initials of the dictator and the secretary should be placed flush with the left-hand margin, two spaces below the last line of the signature. Capitals are usually used for the dictator's initials, and lowercase letters are usually used for the typist's initials.

The following are the various accepted forms for typing initials:

HD/ca   HD:ca   HDca

● *LETTER STYLE*

A competent secretary or stenographer gives careful attention to the mechanical setup of the letter—the arrangement on the page, the width of the margins, and the position of the date, the inside address, the salutation, and the complimentary close. Modern business practice has established certain conventions, and observance of these conventions reflects an efficient office.

The Format

Selecting an appropriate format for your business letters is good business policy, since it adds to the overall impression of you and your company. Therefore, make sure that your secretary uses the best format for your purpose.

Style of type.   When your secretary is planning the layout of a letter, he or she must know the approximate number of words that will fit comfortably on a page. This is partly determined by the kind of type on the machine he or she uses.

The most popular forms of type (whether the secretary uses word processing or electric typewriter) are called "pica" and "elite." (There are other types of styles that you can order, if you wish.) Pica type produces 10 characters per inch across a page, or 85 characters across a sheet of standard size letter paper, 8½" by 11".

The elite style of type is smaller than the pica, and permits 12 characters per inch across a page, or 102 characters across an 8½" by 11" sheet.

PRENTICE HALL
Englewood Cliffs, NJ 07632

                              July 16, 19—
                              Your reference 12:-3:1

Ms Sheila Jones
12 Harrington Place
Greenpoint, N.Y. 00000

Dear Ms Jones:

You asked me if there is any one style of setting up a
letter that is used more than the others. Probably more
business concerns use the block style of letter than any
other style, because its marginal uniformity saves time
for the typist. This letter is an example of the block
style.

As you can see, the inside address is blocked and the
paragraph beginnings are aligned with the left margin.
Open punctuation is used in the address.

The date and reference lines are flush with the right
margin. The date line is two spaces below the letterhead, and the reference line is two spaces below the
date line. The complimentary close begins slightly to
the right of the center of the page. Both lines of the
signature are aligned with the complimentary close.

As the dictator's name is typed in the signature, it is
not considered necessary to include his or her initials
in the identification line.

                              Sincerely yours,

                              Martha Scott
                              Correspondence Chief

cf

(The distinguishing feature of this letter is that it contains a reference line and that the
inside address and paragraphs are blocked, flush with the left-hand margin. Open punctuation is used in the address and signature, and the dictator's initials are not included
in the identification line.)

PRENTICE HALL
Englewood Cliffs, NJ 07632

July 16, 19—

Ms Sheila Jones
12 Harrington Place
Greenpoint, New York 00000

Dear Ms Jones:

                    Subject: Business Letter Styles

   Most companies have a definite preference as to letter
style. Many leading business corporations insist that
all letters be typed in semiblock style. This style
combines an attractive appearance with utility. Private
secretaries, who are not usually concerned with mass
production of correspondence, favor it.

   This style differs from the block form in that first
line of each paragraph is indented five or ten spaces.
As in all letters, there is a double space between paragraphs.

   The date line is flush with the left-hand margin, two
or four spaces below the letterhead. The complimentary
close begins slightly to the left of the center of the
page. All lines of the signature are aligned with the
complimentary close.

                    Very sincerely yours,

                    Martha Scott
                    Correspondence Chief

MS/cf

(The distinguishing feature of a semiblock style of letter is that all structural parts of the
letter begin flush with the left-hand margin, but the first line of each paragraph is indented
five or ten spaces. Open punctuation is used in the address and signature, and a subject
line is included.)

PRENTICE HALL
Englewood Cliffs, NJ 07632

July 16, 19—

Ms Sheila Jones,
   The Modern School for Secretaries,
      12 Harrington Place,
         Greenpoint, New York 00000

Dear Ms Jones:

This is an example of the indented style of letter which many conservative organizations still use. The indented style is correct, however, for any type of firm.

Each line of the address is indented five spaces more than the preceding line. The beginning of each paragraph is indented the same as the third line of the address, which is ten spaces. The complimentary close begins a few spaces to the right of the center of the page, and the lines of the signature are aligned with the complimentary close. Closed punctuation is used in the address but not in the signature.

Very truly yours,

Martha Scott
Correspondence Chief

Ms:cf
Enc.

(The distinguishing feature of the indented style of letter is that each line of address is indented more than the preceding line. The first line of each paragraph is also indented. Close punctuation is used in the address but not in the signature.)

PRENTICE HALL
Englewood Cliffs, NJ 07632

                                        July 16, 19—

Dear Ms Jones:

   Every correspondence manual should include a sample of
the official style. It is used by executives and professional individuals when writing personal letters, and it
looks very well on the executive-size letterhead.

   The structural parts of the letter differ from the
standard arrangement only in the position of the inside
address. The salutation is placed two to five spaces
below the date line, depending upon the length of the
letter. It establishes the left margin of the letter.
The inside address is written in block form, flush with
the left margin, from two to five spaces below the final
line of the signature. Open punctuation is used in the
address.

   The identification line, if used, is placed two spaces
below the last line of the address.

                                Sincerely yours,

                                Martha Scott
                                Correspondence Chief

Ms Sheila Jones
12 Harrington Place
Greenpoint, New York

MS/cf

(The distinguishing feature of the official style of letter is that the inside address is placed below the signature, flush with the left-hand margin, instead of before the salutation. The identification line and enclosure notations, if any, are typed two spaces below the last line of the address. Open punctuation is used.)

PRENTICE HALL
Englewood Cliffs, NJ 07632

                                    July 16, 19—

Ms Sheila Jones
12 Harrington Place
Greenpoint, New York 00000

Dear Ms Jones:

   This is an example of a short letter. The style differs from the previous sample letters in that the lines
are double-spaced and the beginning of each paragraph is
indented five or ten spaces.

   As you can see, the date is typed in the conventional
position, and the complimentary close and the signature
below it start a few spaces to the right of the center
of the page.

                              Very truly yours,

                              Martha Scott
                              Correspondence Chief

MS:cf

(The distinguishing feature of a short letter is that though the body of the letter is double-
spaced for better appearance, the inside address, which uses open punctuation, is still
single-spaced. Each paragraph is indented five or ten spaces to indicate a new paragraph.)

Margins.   A letter can't present a balanced, attractive appearance unless the typist selects the right margin width for the length of the letter. An experienced typist can do this by just glancing at his or her steno book. Suppose, for example, that a particular letter contains 200 words. The address for a letter of this length should begin about 3″ from the top of the page. The proper place for the complimentary close and signature would be about 3½″ from the bottom of the 8½″ by 11″ sheet.

The margins should be set up so that the top and bottom and both sides form a balanced arrangement. In general, the shorter the letter, the wider the margin it needs.

Spacing.   Single-spaced letter with double-spacing between paragraphs is the form most widely used in business correspondence.

Common sense can be your best guide in determining how a letter should be spaced. If a letter contains so few words that a single-spaced treatment would seem to make the words stand lonesome on a page, double-spacing would be called for (or the use of smaller stationery). Generally, letters of 100 words or less look best double-spaced.

The format should be pleasing to the eye. When it comes to deciding how the various parts of a business letter should be arranged on a page, there is no one style that might be considered "the best." The choice of form to be used is usually a matter of preference. Many business concerns and executives have standards for letter styles. On pages 000 through 000 are examples of the more popular formats that can be used as a guide in determining which best suits your purpose and taste.

Indention or tabulation.   Indention is a device that creates a "frame within a frame." It is especially effective when the letter is typed in the block form. Because the left-hand margin of the block letter is so strictly established, the indention of certain material stands out particularly well—in the same way that a "mat" serves to set off a picture within its initial frame.

example

The envelope we're enclosing is to make it easy for you to turn this sheet over and write us—

to have our representative call,
to send you our price list,
or, to say just what you think of us.

We hope it's good, but if it isn't . . .

Numbering. Numbering is used effectively to set off various points that are later detailed. The numbered items may be listed one beneath the other, or they may be run together within a paragraph. If the latter method is used, the numbers should be enclosed in parentheses, for example, (1). If the listing method is used, the following procedure is advised.

1. Indent five spaces from each margin of the letter—more if necessary to center the material.

2. Precede each item with a number, followed by a period, for example,

> 4. Quality Control

3. Begin each line of the indented material two spaces to the right of the number.

4. Single space the material within each item, but double space between items, for example,

> 4. Quality Control—in
>   Manufacturing
>   Packaging
>   Shipping
>
> 5. Security—in
>   Physical access to the plant
>   Authorized personnel to the
>   design room

Sometimes the typist will want to use both the listing and the run-in methods in a single letter, especially when the subject matter requires subdivision within the original numbering of facts.

Underscoring and capitalization. A line under an important phrase or topical heading is used for emphasis in presenting facts. Another device for attracting the reader's attention is the capitalization of a phrase or heading. But whether they are used individually or together, both are seasonings that must be used with caution. Too much underscoring and capitalization makes the reader feel that much ado is being made about nothing. It distracts his or her attention, rather than captures it.

Forms of punctuation. The punctuation of the heading, inside address, and signature may be either open or close, as the previous sample letters displayed. But whichever style is selected for one part of the letter, must be used for all three. Again, consistency of style can make or break the appearance of a letter. Hence, the punctuation of the heading sets the style for the rest of the letter.

1. *Open punctuation* means the omission of all commas and periods at the ends of lines in the heading, the inside address, and the signature.
2. *Close (or closed) punctuation,* no longer popular, requires that punctuation be used after each line in the heading, the inside address, and the signature—if the typed name of the letter writer is followed by his or her official title.

Here's an example of each of the two styles of punctuation:

open

June 19, 19—

Mr. William J. Johnson
97 Arden Street
Philadelphia, PA 00000

Dear Mr. Johnson:

> Sincerely,
> (signature)
> Nancy Jorgensen
> Vice-President

closed

June 19, 19—

Mr. William J. Johnson,
97 Arden Street,
Philadelphia, PA. 00000

Dear Mr. Johnson:

> Sincerely,
> (signature)
> Nancy Jorgensen,
> Vice-President

It's important to remember that the salutation is usually followed by a colon and the complimentary close by a comma, no matter which of the two punctuation forms is used for the other parts of the letter.

To make the following section as useful as possible for the busy professional whose many responsibilities include writing all types of letters, it is organized as a reference guide in alphabetical order.

Proper Address Form

For the overall impression of a letter it is quite important that the address form is proper. For formal letters, the charts in Chapter 11 provide the correct titles of U.S. and foreign officials. Proper address form for less formal letters, however, is just as important. This is especially true when the letter goes to a woman executive. When the letter goes to a male executive, the address form is rather simple. For example,

> Mr. Robert Reed
> Vice-President, Engineering
> Brady Corporation
> 14 Waverly Place
> Houston, Texas 00000

But, when the letter goes to a female executive, the title can be "Miss," "Mrs." or "Ms" (which, by the way, is *not* an abbreviation, and consequently does not require a period after it just as "Miss" does not). To play it safe, more and more executives use the "Ms," which is an appropriate form of address to either a single or a married woman. For example,

> Ms Jennifer Reed
> Vice-President, Technical Services
> Brady Corporation
> Waverly Place
> Houston, Texas 00000

The latest trend in the business world is to go one step farther and leave out the title in the address form entirely. For example,

> Robert Reed
> Vice-President, Marketing
> and
> Jennifer Reed
> Vice-President, Marketing

Complimentary Close

The complimentary close, which is typed two spaces below the last paragraph of the letter, should correspond with the salutation both in form and the degree of formality. In other words, if you began your letter with the very formal "My dear Sir:" or "Ladies:", you wouldn't use "Cordially yours," as a complimentary close. Similarly, if your salutation reads "Dear Henry:", you wouldn't use the complimentary close "Respectfully yours."

First Sentence and Opening Paragraph

With your first sentence, first paragraph, you should indicate not only the subject of your letter but establish its mood as well.

Your opening sentence can be complimentary, gracious, provocative, challenging, or a statement of facts—depending on the subject of your letter and the impact you wish to have on the addressee. The main thing to remember is that it is your first sentence that prepares the reader for the rest of your communication. Here, for example, is the first sentence of a letter that is bound to attract the reader's attention, if he or she is thinking of a vacation trip.

If you are planning to take a vacation trip this year, we have a suggestion which can make this the best trip you ever made—and can save you money too.

For another example, here's a rather provocative opening of a different type of letter:

The bird is the fastest moving living creature. Speeds higher than 185 mph have been attributed to peregrine falcons. Homo sapiens, of course, move less quickly. But there is one group which is known for its remarkable mobility—the business executive.[1]

For still another example, here's the first sentence, first paragraph of a letter that intends to relay harsh realities to the reader.

[1] Through the courtesy of *Fortune* © 1979, Time Inc.

The premium of 2 percent cash discount is for early payment. Since we did not receive your check for our invoice #12576 until 30 days after the discount period expired, we cannot allow your deduction.

And finally, here's an example of the beginning of a letter that combines criticism with graciousness.

In a recent commentary on "Disciplining Science Teachers," your associate editor observed that criticizing schools and complaining about teachers seems to be "increasingly fashionable." He is uncomfortably correct. But, may I also say, that it's a myth that our technical schools are failing.

Format

The primary consideration in deciding upon a format is that it should be pleasing to the eye. Actually, when it comes to determining how the various parts of a business letter should be arranged on a page, there is no one style that might be considered "the best." The choice of form to be used is usually a matter of preference. A letter can be in a block form or indented (the first line of each paragraph indented five or ten spaces), but as long as there's a nice wide margin all around the single-spaced letter, it will give a balanced, attractive appearance. If the letter is quite short, it will look better if it is double-spaced. Moreover, a double-spaced letter should use the indented form to indicate new paragraphs.

The Last Sentence

The last sentence should reinforce the first sentence and the topic of the letter. It should not jar the reader. Thus, for example, if it's a letter politely reminding the client of an outstanding bill, the last sentence cannot suddenly be a sharp one. Conversely, if the letter is a firm, no-nonsense rejection of an offer or an applicant, the last sentence should not be, abruptly, a really warm one. In short, the last sentence should be in tune with the rest of the letter and leave the reader with a clear understanding of the subject and the intent of the letter. For example, here's the last sentence of a letter written to a vacation-bound person.

It would be our pleasure to help you make this the most interesting, relaxed, and economical trip you ever had.

For another example, here's the last paragraph of a letter for "Contacting Inactive Accounts":

Won't you please let me know personally why you have stopped placing orders with us? I would greatly appreciate it.

As a final example, here's the last sentence of a letter for "Demanding Retraction":

We are looking forward to a retraction of the above item as soon as possible in your newspaper.

Letterhead

Quality paper enhances the effectiveness of a business letterhead. But the best quality paper won't help if the printed letterhead is offensively ostentatious. The most impressive business letterheads are simple and subdued. For example, a simple brown letterhead printed on ivory or light-tan quality paper gives the impression of a company that knows its own worth and doesn't need to shout it from the rooftops. That doesn't mean, however, that a bold, black letterhead printed on white paper cannot be just as tasteful and impressive. The fact is that the choice of printed letterhead is entirely a matter of taste and preference. The main thing is that it reflects the personality of the company and the particular field in which the company is operating.

Paper

Quality paper can be judged by its bulk, by its ability to reproduce the typewritten material clearly, and by the way it can withstand erasures. The most commonly used paper for business letters is the 20-pound weight paper, because it satisfies all the foregoing criteria.

Paragraphs

To make your letters interesting, you should vary the length of your paragraphs. In fact, a one-sentence or even a one-word paragraph can sparkle the interest of the most indifferent addressee. Of course, an explanatory paragraph enlarging upon the eye-catching exclamation should follow. For example, here are the first two paragraphs from a letter written to a "nonactive" client account.

Tell us what you think!

As a valued client, we'd like your opinion as to how we could improve our services.

For another example, here are the first two paragraphs from a letter written by an airline company.

Our apologies!

I sincerely regret that we were unable to operate your August 3 flight on time, and apologize for the inconvenience it caused you.

For still another example, here are the first two paragraphs from a letter soliciting for charitable contribution.

Help!

We really need your help. The work load of Hope Indian School has tripled in the past year due to increased enrollment. As a result, so much more is asked of us.

Salutation

The form of salutation depends on the relationship between you and the addressee as well as the tone of the letter. If it's a formal business letter addressed to a man, make the salutation singular, for example, "Dear Mr. Silvers," or "Dear Sir," if you want to be really formal. If the letter is addressed to a company or group of men, make it plural, for example, "Gentlemen," or "Dear Sirs."

If the formal business letter is addressed to a woman, make the salutation singular, for example, "Dear Ms Silvers," or "Dear Madam," if you want to be really formal. If the letter is addressed to a company or group of women, make it plural, for example, "Ladies," or "Mesdames."

Of course, if you know the addressee very well, even if it's a business letter, it's appropriate to make the salutation "Dear Jim," or "Dear Jennifer."

Signature

Your signature on a business letter usually consists of your written signature above your typed name and business title, and the name of your division or department, if appropriate. Business titles and degree letters follow the typed signature. For example, *Paul Steward, President, Patricia Wolner, Ph.D.,* or *James Fall, Director.* No title (*Mr. or Ms* or *Dr. or Professor*) precedes either the written or the typed signature.

Style

The style of your letter and the style of your writing should be more than just gracious and correct, it should reflect your personality. Consequently, whether it's a formal or informal letter that you are writing or dictating, don't use stiff, pale words. Try to inject colorful, active verbs that will let your personality shine through and that will make reading your letters a pleasure.

Tone

Just as the tone of your voice will reflect the mood you're in, the tone of your letter will mirror your attitude and your thoughts. Consequently, if you have a condescending attitude toward the person you're writing to, the tone of your letter will reflect that, no matter how hard you're trying to hide it behind "nice" words. Similarly, artificial geniality or warmth is easily detected in a letter, regardless of how careful the writer was in composing the letter. Therefore, the best way to ensure that the tone of your letter doesn't strike some "sour notes" is for you to be sincere. The tone of a letter written with sincerity will always be well received.

A WORD ON STYLE

Do Not Overuse I

When you write, you have a natural tendency to use the word "I". You have to refer to yourself in some way, and neither the imperial "we" nor the third person is ordinarily a good choice. Is there really anything wrong with I? No—not if it is used in moderation.

The problem with I arises when it is overused. If every paragraph in a five-paragraph letter begins with I, for instance, the reader is likely to be put off, consciously or subconsciously. On the other hand, if only one of the paragraphs, preferably not the first one, begins with I, no reasonable reader should object.

Chapter 11

REFERENCE GUIDE TO PROPER ADDRESS FORMS

U.S. Government Officials . . . State and Local Government
Officials . . . Court Officials . . . U.S. Diplomatic
Representatives . . . Foreign Officials and
Representatives . . . The U.S. Armed Forces: The
Army . . . The Navy . . . The Air Force . . . The Marine
Corps . . . The Coast Guard . . . Religious Dignitaries:
Catholic faith . . . Jewish faith . . . Protestant
faith . . . College and University Officials . . . The United
Nations . . . The Organization of the American
States . . . Miscellaneous . . . British Forms: The peerage,
baronets, and knights . . . British government officials

Correct use of the proper address form is an essential part of your letter. Consequently, just as you would not tolerate a letter leaving your office with poor grammar, misspelled words, or sloppy erasures, you would not condone a casual approach to forms of address. The reason is a practical one: should you offend your reader by addressing him or her incorrectly, you might completely nullify the effect of an otherwise excellent letter. It is an unusual person who does not take his or her title seriously, regardless of title or rank.

The useful data that follows will save you substantial time and effort when you write to an individual with an official title. Each chart provides you with the correct forms of written address, salutation, and complimentary close to be used in letters to titled persons. This handy reference guide will serve you well on these important occasions.

Correct Forms of Address

UNITED STATES GOVERNMENT OFFICIALS

Personage	Envelope and Inside Address (Add City, State, Zip)	Formal Salutation	Informal Salutation	Formal Close	Informal Close	Spoken Address / Informal Introduction or Reference
The President	The President The White House	Mr. President	Dear Mr. President:	Respectfully yours,	Very respectfully yours, *or* Sincerely yours,	1. Mr. President 2. Not introduced (The President)
Former President of the United States[1]	The Honorable William R. Blank (local address)	Sir:	Dear Mr. Blank:	Respectfully yours,	Sincerely yours,	1. Mr. Blank 2. Former President Blank *or* Mr. Blank
The Vice-President of the United States	The Vice-President of the United States United States Senate	Mr. Vice-President:	Dear Mr. Vice-President:	Very truly yours,	Sincerely yours,	1. Mr. Vice-President *or* Mr. Blank The Vice-President
The Chief Justice of the United States Supreme Court	The Chief Justice of the United States The Supreme Court of the United States	Sir:	Dear Mr. Chief Justice:	Very truly yours,	Sincerely yours,	1. Mr. Chief Justice 2. The Chief Justice
Associate Justice of the United States Supreme Court	Mr. Justice Blank The Supreme Court of the United States	Sir:	Dear Mr. Justice:	Very truly yours,	Sincerely yours,	1. Mr. Justice Blank *or* Justice Blank 2. Mr. Justice Blank

1. If a former president has a title, such as *General of the Army*, address him by it.

	Address	Formal Salutation	Informal Salutation	Formal Closing	Informal Closing	Informal Salutation / Reference
Retired Justice of the United States Supreme Court	The Honorable William R. Blank (local address)	Sir:	Dear Justice Blank:	Very truly yours,	Sincerely yours,	1. Mr. Justice Blank *or* Justice Blank 2. Mr. Justice Blank
The Speaker of the House of Representatives	The Honorable William R. Blank Speaker of the House of Representatives	Sir:	Dear Mr. Speaker: *or* Dear Mr. Blank:	Very truly yours,	Sincerely yours,	1. Mr. Speaker *or* Mr. Blank 2. The Speaker, Mr. Blank (The Speaker *or* Mr. Blank)
Former Speaker of the House of Representatives	The Honorable William R. Blank (local address)	Sir:	Dear Mr. Blank:	Very truly yours,	Sincerely yours,	1. Mr. Blank 2. Mr. Blank
Cabinet Officers addressed as "Secretary"[2]	The Honorable William R. Blank Secretary of State The Honorable William R. Blank Secretary of State of the United States of America (if written from abroad)	Sir:	Dear Mr. Secretary:	Very truly yours,	Sincerely yours,	1. Mr. Secretary *or* Secretary Blank *or* Mr. Blank 2. The Secretary of State Mr. Blank (Mr. Blank *or* The Secretary)
Former Cabinet Officer	The Honorable William R. Blank (local address)	Dear Sir:	Dear Mr. Blank:	Very truly yours,	Sincerely yours,	1. Mr. Blank 2. Mr. Blank
Postmaster General	The Honorable William R. Blank The Postmaster General,	Sir:	Dear Mr. Postmaster General:	Very truly yours,	Sincerely yours,	1. Mr. Postmaster General *or* Postmaster General Blank *or* Mr. Blank 2. The Postmaster General, Mr. Blank (Mr. Blank *or* The Postmaster General)

2. Titles for cabinet secretaries are Secretary of State; Secretary of the Treasury; Secretary of Defense; Secretary of the Interior; Secretary of Agriculture; Secretary of Commerce; Secretary of Labor; Secretary of Health and Human Services; Secretary of Housing and Urban Development; Secretary of Transportion.

UNITED STATES GOVERNMENT OFFICIALS *continued*

Personage	Envelope and Inside Address (Add City, State, Zip)	Formal Salutation	Informal Salutation	Formal Close	Informal Close	1. Spoken Address 2. Informal Introduction or Reference
The Attorney General	The Honorable William R. Blank The Attorney General	Sir:	Dear Mr. Attorney General:	Very truly yours,	Sincerely yours,	1. Mr. Attorney General *or* Attorney General Blank 2. The Attorney General, Mr. Blank (Mr. Blank *or* The Attorney General)
Under Secretary of a Department	The Honorable William R. Blank Under Secretary of Labor	Dear Mr. Blank:	Dear Mr. Blank:	Very truly yours,	Sincerely yours,	1. Mr. Blank 2. Mr. Blank
United States Senator	The Honorable William R. Blank United States Senate	Sir:	Dear Senator Blank:	Very truly yours,	Sincerely yours,	1. Senator Blank *or* Senator 2. Senator Blank
Former Senator	The Honorable William R. Blank (local address)	Dear Sir:	Dear Senator Blank:	Very truly yours,	Sincerely yours,	1. Senator Blank *or* Senator 2. Senator Blank
Senator-elect	Honorable William R. Blank Senator-elect United States Senate	Dear Sir:	Dear Mr. Blank:	Very truly yours,	Sincerely yours,	1. Mr. Blank 2. Senator-elect Blank *or* Mr. Blank
Committee Chairman— United States Senate	The Honorable William R. Blank, Chairman Committee on Foreign Affairs United States Senate	Dear Mr. Chairman:	Dear Mr. Chairman: *or* Dear Senator Blank:	Very truly yours,	Sincerely yours,	1. Mr. Chairman *or* Senator Blank *or* Senator 2. The Chairman *or* Senator Blank

	Address	Formal Salutation	Informal Salutation	Formal Close	Informal Close	Formal / Informal Address
Subcommittee Chairman— United States Senate	The Honorable William R. Blank, Chairman, Subcommittee on Foreign Affairs United States Senate	Dear Senator Blank;	Dear Senator Blank:	Very truly yours,	Sincerely yours.	1. Senator Blank *or* Senator 2. Senator Blank
United States Representative or Congressman[3]	The Honorable William R. Blank House of Representatives The Honorable William R. Blank Representative in Congress (local address) (when away from Washington, DC)	Sir:	Dear Mr. Blank:	Very truly yours,	Sincerely yours,	1. Mr. Blank 2. Mr. Blank, Representative (Congressman) from New York *or* Mr. Blank
Former Representative	The Honorable William R. Blank (local address)	Dear Sir: *or* Dear Mr. Blank:	Dear Mr. Blank:	Very truly yours,	Sincerely yours,	1. Mr. Blank 2. Mr. Blank
Territorial Delegate	The Honorable William R. Blank Delegate of Puerto Rico House of Representatives	Dear Sir: *or* Dear Mr. Blank:	Dear Mr. Blank;	Very truly yours,	Sincerely yours,	1. Mr. Blank 2. Mr. Blank
Resident Commissioner	The Honorable William R. Blank Resident Commissioner of (Territory) House of Representatives	Dear Sir: *or* Dear Mr. Blank:	Dear Mr. Blank:	Very truly yours,	Sincerely yours,	1. Mr. Blank 2. Mr. Blank
Directors or Heads of Independent Federal Offices, Agencies Commissions, Organizations, etc.	The Honorable William R. Blank Director, Mutual Security Agency	Dear Mr. Director (Commissioner, etc.):	Dear Mr. Blank:	Very truly yours,	Sincerely yours,	1. Mr. Blank 2. Mr. Blank

3. The official title of a "congressman" is *Representative.* Senators are also congressmen.

UNITED STATES GOVERNMENT OFFICIALS *continued*

Personage	Envelope and Inside Address (Add City, State, Zip)	Formal Salutation	Informal Salutation	Formal Close	Informal Close	1. Spoken Address 2. Informal Introduction or Reference
Other High Officials of the United States, in general: Public Printer, Comptroller General	The Honorable William R. Blank Public Printer The Honorable William R. Blank Comptroller General of the United States	Dear Sir: *or* Dear Mr. Blank:	Dear Mr. Blank:	Very truly yours,	Sincerely yours,	1. Mr. Blank 2. Mr. Blank

Personage*	Envelope and Inside Address (Add City, State, Zip)	Formal Salutation	Informal Salutation	Formal Close	Informal Close	1. Spoken Address 2. Informal Introduction or Reference
Secretary to the President	The Honorable William R. Blank Secretary to the President The White House	Dear Sir: *or* Dear Mr. Blank:	Dear Mr. Blank:	Very truly yours,	Sincerely yours,	1. Mr. Blank 2. Mr. Blank
Assistant Secretary to the President	The Honorable William R. Blank Assistant Secretary to the President The White House	Dear Sir: *or* Dear Mr. Blank:	Dear Mr. Blank:	Very truly yours,	Sincerely yours,	1. Mr. Blank 2. Mr. Blank
Press Secretary to the President	Mr. William R. Blank Press Secretary to the President The White House	Dear Sir: *or* Dear Mr. Blank:	Dear Mr. Blank:	Very truly yours,	Sincerely yours,	1. Mr. Blank 2. Mr. Blank

STATE AND LOCAL GOVERNMENT OFFICIALS

	Address	Salutation	Salutation	Complimentary Close	Complimentary Close	In Speaking / Introducing
Governor of a State or Territory[1]	The Honorable William R. Blank, Governor of New York	Sir:	Dear Governor Blank:	Very truly yours,	Sincerely yours,	1. Governor Blank *or* Governor 2. a) Governor Blank b) The Governor c) The Governor of New York (used only outside his or her own state)
Acting Governor of a State or Territory	The Honorable William R. Blank, Acting Governor of Connecticut	Sir:	Dear Mr. Blank:	Very truly yours,	Sincerely yours,	1. Mr. Blank 2. Mr. Blank
Lieutenant Governor	The Honorable William R. Blank, Lieutenant Governor of Iowa	Sir:	Dear Mr. Blank:	Very truly yours,	Sincerely yours,	1. Mr. Blank 2. The Lieutenant Governor of Iowa, Mr. Blank *or* The Lieutenant Governor
Secretary of State	The Honorable William R. Blank, Secretary of State of New York	Sir:	Dear Mr. Secretary:	Very truly yours,	Sincerely yours,	1. Mr. Blank 2. Mr. Blank
Attorney General	The Honorable William R. Blank, Attorney General of Massachusetts	Sir:	Dear Mr. Attorney General:	Very truly yours,	Sincerely yours,	1. Mr. Blank 2. Mr. Blank
President of the Senate of a State	The Honorable William R. Blank, President of the Senate of the State of Virginia	Sir:	Dear Mr. Blank:	Very truly yours,	Sincerely yours,	1. Mr. Blank 2. Mr. Blank

1. The form of addressing governors varies in the different states. The form given here is the one used in most states. In Massachusetts by law and in some other states by courtesy, the form is *His (Her) Excellency, the Governor of Massachusetts.*

STATE AND LOCAL GOVERNMENT OFFICIALS *continued*

Personage	Envelope and Inside Address (Add City, State, Zip)	Formal Salutation	Informal Salutation	Formal Close	Informal Close	1. Spoken Address 2. Informal Introduction or Reference
Speaker of the Assembly or The House of Representatives. [2]	The Honorable William R. Blank Speaker of the Assembly of the State of New York	Sir:	Dear Mr. Blank:	Very truly yours,	Sincerely yours,	1. Mr. Blank 2. Mr. Blank
Treasurer, Auditor, or Comptroller of a State	The Honorable William R. Blank Treasurer of the State of Tennessee	Dear Sir:	Dear Mr. Blank:	Very truly yours,	Sincerely yours,	1. Mr. Blank 2. Mr. Blank
State Senator	The Honorable William R. Blank The State Senate	Dear Sir:	Dear Senator Blank:	Very truly yours,	Sincerely yours,	1. Senator Blank *or* Senator 2. Senator Blank
State Representative, Assemblyman, or Delegate	The Honorable William R. Blank House of Delegates	Dear Sir:	Dear Mr. Blank:	Very truly yours,	Sincerely yours,	1. Mr. Blank 2. Mr. Blank *or* Delegate Blank
District Attorney	The Honorable William R. Blank District Attorney, Albany County County Courthouse	Dear Sir:	Dear Mr. Blank:	Very truly yours,	Sincerely yours,	1. Mr. Blank 2. Mr. Blank
Mayor of a city	The Honorable William R. Blank Mayor of Detroit	Dear Sir:	Dear Mayor Blank:	Very truly yours,	Sincerely yours,	1. Mayor Blank *or* Mr. Mayor 2. Mayor Blank
President of a Board of Commissioners	The Honorable William R. Blank, President Board of Commissioners of the City of Buffalo	Dear Sir:	Dear Mr. Blank:	Very truly yours,	Sincerely yours,	1. Mr. Blank 2. Mr. Blank

2. In most states the lower branch of the legislature is the House of Representatives. The exceptions to this are: New York, California, Wisconsin, and Nevada, where it is known as the Assembly; Maryland, Virginia, and West Virginia—the House of Delegates; New Jersey—the House of General Assembly.

City Attorney, City Counsel Corporation Counsel	The Honorable William R. Blank, City Attorney (City Counsel, Corporation Counsel)	Dear Sir:	Dear Mr. Blank:	Very truly yours,	Sincerely yours,	1. Mr. Blank 2. Mr. Blank
Alderman	Alderman William R. Blank City Hall	Dear Sir:	Dear Mr. Blank:	Very truly yours,	Sincerely yours,	1. Mr. Blank 2. Mr. Blank

COURT OFFICIALS

Chief Justice[1] of a State Supreme Court	The Honorable William R. Blank Chief Justice of the Supreme Court of Minnesota[2]	Sir:	Dear Mr. Chief Justice:	Very truly yours,	Sincerely yours,	1. Mr. Chief Justice *or* Judge Blank 2. Mr. Chief Justice Blank *or* Judge Blank
Associate Justice of a Supreme Court of a State	The Honorable William R. Blank Associate Justice of the Supreme Court of Minnesota	Sir:	Dear Justice Blank:	Very truly yours,	Sincerely yours,	1. Mr. Justice Blank 2. Mr. Justice Blank
Presiding Justice	The Honorable William R. Blank Presiding Justice, Appellate Division Supreme Court of New York	Sir:	Dear Justice Blank:	Very truly yours,	Sincerely yours,	1. Mr. Justice (*or* Judge) Blank 2. Mr. Justice (*or* Judge) Blank
Judge of a Court[3]	The Honorable William R. Blank Judge of the United States District Court for the Southern District of California	Sir:	Dear Judge Blank:	Very truly yours,	Sincerely yours,	1. Judge Blank 2. Judge Blank

1. If his or her official title is *Chief Justice* substitute *Chief Judge* for *Chief Justice*, but never use *Mr.*, *Mrs.*, *Miss*, or *Ms.* with *Chief Judge* or *Judge*.

2. Substitute here the appropriate name of the court. For example, the highest court in New York State is called the Court of Appeals.

3. Not applicable to judges of the United States Supreme Court:

Personage	Envelope and Inside Address (Add City, State, Zip or City, Country)	Formal Salutation	Informal Salutation	Formal Close	Informal Close	1. Spoken Address 2. Informal Introduction or Reference
Clerk of a Court	William R. Blank, Esq. Clerk of the Superior Court of Massachusetts	Dear Sir:	Dear Mr. Blank:	Very truly yours,	Sincerely yours,	1. Mr. Blank 2. Mr. Blank

UNITED STATES DIPLOMATIC REPRESENTATIVES

Personage	Envelope and Inside Address (Add City, State, Zip or City, Country)	Formal Salutation	Informal Salutation	Formal Close	Informal Close	1. Spoken Address 2. Informal Introduction or Reference
American Ambassador	The Honorable William R. Blank American Ambassador[1]	Sir:	Dear Mr. Ambassador:	Very truly yours,	Sincerely yours,	1. Mr. Ambassador *or* Mr. Blank 2. The American Ambassador[2] (The Ambassador *or* Mr. Blank)
American Minister	The Honorable William R. Blank American Minister to Rumania	Sir:	Dear Mr. Minister:	Very truly yours,	Sincerely yours,	1. Mr. Minister *or* Mr. Blank 2. The American Minister, Mr. Blank (The Minister *or* Mr. Blank)
American Chargé d' Affaires, Consul General, Consul, or Vice Consul	William R. Blank, Esq. American Chargé d'Affaires ad interim (or other title)	Sir:	Dear Mr. Blank:	Very truly yours,	Sincerely yours,	1. Mr. Blank 2. Mr. Blank
High Commissioner	The Honorable William R. Blank United States High Commissioner to Argentina	Sir:	Dear Mr. Blank:	Very truly yours,	Sincerely yours,	1. Commissioner Blank *or* Mr. Blank 2. Commissioner Blank *or* Mr. Blank

1. When an ambassador or minister is not at his or her post, the name of the country to which he or she is accredited must be added to the address. For example: *The American Ambassador to Great Britain.* If he or she holds military rank, the diplomatic complimentary title *The Honorable* should be omitted, thus *General William R. Blank, American Ambassador (or Minister).*

2. With reference to ambassadors and ministers to Central or South American countries, substitute *The Ambassador of the United States* for *American Ambassador* or *American Minister.*

FOREIGN OFFICIALS AND REPRESENTATIVES

Foreign Ambassador[1] in the United States	His Excellency,[2] Erik Rolf Blankson Ambassador of Norway	Excellency:	Dear Mr. Ambassador:	Very truly yours,	Sincerely yours,	1. Mr. Ambassador *or* Mr. Blank 2. The Ambassador of Norway (The Ambassador *or* Mr. Blank)
Foreign Minister[3] in the United States	The Honorable George Macovescu Minister of Rumania	Sir:	Dear Mr. Minister:	Very truly yours,	Sincerely yours,	1. Mr. Minister *or* Mr. Blank 2. The Minister of Rumania (The Minister *or* Mr. Blank)
Foreign Diplomatic Representative with a Personal Title[4]	His Excellency,[5] Count Allesandro de Bianco Ambassador of Italy	Excellency:	Dear Mr. Ambassador:	Very truly yours,	Sincerely yours,	1. Mr. Ambassador *or* Count Bianco 2. The Ambassador of Italy (The Ambassador *or* Count Bianco)
Prime Minister	His Excellency, Christian Jawaharal Blank Prime Minister of India	Excellency:	Dear Mr. Prime Minister:	Respectfully yours,	Sincerely yours,	1. Mr. Blank 2. Mr. Blank *or* The Prime Minister
British Prime Minister	The Right Honorable Godfrey Blank, K.G., M.C., M.P. Prime Minister	Sir:	Dear Mr. Prime Minister: *or* Dear Mr. Blank:	Respectfully yours,	Sincerely yours,	1. Mr. Blanc 2. Mr. Blanc *or* The Prime Minister

1. The correct title of all ambassadors and ministers of foreign countries is *Ambassador (Minister) of _____ (name of country)*, with the exception of Great Britain. The adjective form is used with reference to representatives from Great Britain—*British Ambassador, British Minister*.

2. When the representative is British or a member of the British Commonwealth, it is customary to use *The Right Honorable* and *The Honorable* in addition to *His (Her) Excellency*, wherever appropriate.

3. The correct title of all ambassadors and ministers of foreign countries is *Ambassador (Minister) of _____ (name of country)*, with the exception of Great Britain. The adjective form is used with reference to representatives from Great Britain—*British Ambassador, British Minister*.

4. If the personal title is a royal title, such as *His (Her) Highness* or *Prince*, the diplomatic title *His (Her) Excellency* or *The Honorable* is omitted.

5. *Dr., Señor Dom*, and other titles of special courtesy in Spanish-speaking countries may be used with the diplomatic title *His (Her) Excellency* or *The Honorable*.

FOREIGN OFFICIALS AND REPRESENTATIVES *continued*

Personage	Envelope and Inside Address (Add City, State, Zip)	Formal Salutation	Informal Salutation	Formal Close	Informal Close	1. Spoken Address 2. Informal Introduction or Reference
Canadian Prime Minister	The Right Honorable Claude Louis St. Blanc, C.M.G. Prime Minister of Canada	Sir:	Dear Mr. Prime Minister: *or* Dear Mr. Blanc:	Respectfully yours,	Sincerely yours,	
President of a Republic	His Excellency, Juan Cuidad Blanco President of the Dominican Republic	Excellency:	Dear Mr. President:	Respectfully yours,	Sincerely yours,	1. Your Excellency 2. Not introduced (President Blanco *or* the President)
Premier	His Excellency, Charles Yves de Blanc Premier of the French Republic	Excellency:	Dear Mr. Premier:	Respectfully yours,	Sincerely yours,	1. Mr. Blanc 2. Mr. Blanc *or* The Premier
Foreign Chargé d'Affaires (de missi)[6] in the United States	Mr. Jan Gustaf Blanc Chargé d'Affaires of Sweden	Sir:	Dear Mr. Blanc:	Very truly yours,	Sincerely yours,	1. Mr. Blanc 2. Mr. Blanc
Foreign Chargé d'Affaires ad interim in the United States	Mr. Edmund Blank Chargé d'Affaires ad interim[7] of Ireland	Sir:	Dear Mr. Blank:	Very truly yours,	Sincerely yours,	1. Mr. Blank 2. Mr. Blank

6. The full title is usually shortened to *Chargé d'Affaires.*
7. The words *ad interim* should not be omitted in the address.

THE ARMED FORCES/THE ARMY

Personage	Envelope and Inside Address (Add City, State, Zip)	Formal Salutation	Informal Salutation	Formal Close	Informal Close	1. Spoken Address 2. Informal Introduction or Reference
General of the Army	General of the Army William R. Blank, U.S.A. Department of the Army	Sir:	Dear General Blank:	Very truly yours,	Sincerely yours,	1. General Blank 2. General Blank
General, Lieutenant General, Major General, Brigadier General	General (Lieutenant General, Major General, or Brigadier General) William R. Blank, U.S.A.[1]	Sir:	Dear General Blank:	Very truly yours,	Sincerely yours,	1. General Blank 2. General Blank
Colonel, Lieutenant Colonel	Colonel (Lieutenant Colonel) William R. Blank, U.S.A.	Dear Colonel Blank:	Dear Colonel Blank:	Very truly yours,	Sincerely yours,	1. Colonel Blank 2. Colonel Blank
Major	Major William R. Blank, U.S.A.	Dear Major Blank:	Dear Major Blank:	Very truly yours,	Sincerely yours,	1. Major Blank 2. Major Blank
Captain	Captain William R. Blank, U.S.A.	Dear Captain Blank:	Dear Captain Blank:	Very truly yours,	Sincerely yours,	1. Captain Blank 2. Captain Blank
First Lieutenant, Second Lieutenant[2]	Lieutenant William R. Blank, U.S.A.	Dear Lieutenant Blank:	Dear Lieutenant Blank:	Very truly yours,	Sincerely yours,	1. Lieutenant Blank 2. Lieutenant Blank
Chief Warrant Officer, Warrant Officer	Mr. William R. Blank, U.S.A.	Dear Mr. Blank:	Dear Mr. Blank:	Very truly yours,	Sincerely yours,	1. Mr. Blank 2. Mr. Blank
Chaplain in the U.S. Army[3]	Chaplain William R. Blank, Captain, U.S.A.	Dear Chaplain Blank:	Dear Chaplain Blank:	Very truly yours,	Sincerely yours,	1. Chaplain Blank 2. Chaplain Blank (Chaplain Blank)

1. *U.S.A.* indicates regular service, *A.U.S.* (Army of the United States) signifies the Reserve.
2. In all *official* correspondence, the full rank should be included in both the envelope and the inside address, but not in the salutation.
3. Roman Catholic chaplains and certain Anglican priests are introduced as *Chaplain Blank* but are spoken to and referred to as *Father Blank*.

THE ARMED FORCES/THE NAVY continued

Personage	Envelope and Inside Address (Add City, State, Zip)	Formal Salutation	Informal Salutation	Formal Close	Informal Close	1. Spoken Address 2. Informal Introduction or Reference
Fleet Admiral	Fleet Admiral William R. Blank, U.S.N. Chief of Naval Operations, Department of the Navy	Sir:	Dear Admiral Blank:	Very truly yours,	Sincerely yours,	1. Admiral Blank 2. Admiral Blank
Admiral, Vice Admiral, Rear Admiral	Admiral (Vice Admiral or Rear Admiral) William R. Blank, U.S.N. United States Naval Academy[1]	Sir:	Dear Admiral Blank:	Very truly yours,	Sincerely yours,	1. Admiral Blank 2. Admiral Blank
Commodore, Captain, Commander, Lieutenant Commander	Commodore (Captain, Commander, Lieutenant Commander) William R. Blank, U.S.N. U.S.S. Mississippi	Dear Commodore (Captain, Commander) Blank:	Dear Commodore (Captain, Commander) Blank:	Very truly yours,	Sincerely yours,	1. Commodore (Captain, Commander) Blank 2. Commodore (Captain, Commander) Blank
Junior Officers: Lieutenant, Lieutenant Junior Grade, Ensign	(Lieutenant, etc.) William R. Blank, U.S.N. U.S.S. Wyoming	Dear Mr. Blank:	Dear Mr. Blank:	Very truly yours,	Sincerely yours,	1. Mr. Blank[2] 2. Lieutenant, etc., Blank (Mr. Blank)
Chief Warrant Officer, Warrant Officer	Mr. William R. Blank, U.S.N. U.S.S. Texas	Dear Mr. Blank:	Dear Mr. Blank:	Very truly yours,	Sincerely yours,	1. Mr. Blank 2. Mr. Blank
Chaplain	Chaplain William R. Blank Captain, U.S.N. Department of the Navy	Dear Chaplain Blank:	Dear Chaplain Blank:	Very truly yours,	Sincerely yours,	1. Chaplain Blank 2. Captain Blank (Chaplain Blank)

1. *U.S.N.* signifies regular service; *U.S.N.R.* indicates the Reserve.
2. Junior officers in the medical or dental corps are spoken to and referred to as *Dr.* but are introduced by their rank.

THE ARMED FORCES—AIR FORCE

Air Force titles are the same as those in the Army. *U.S.A.F.* is used instead of *U.S.A.* and *A.F.U.S.* is used to indicate the Reserve.

THE ARMED FORCES—MARINE CORPS

Marine Corps titles are the same as those in the Army, except that the top rank is *Commandant of the Marine Corps*. *U.S.M.C.* indicates regular service, *U.S.M.R.* indicates the Reserve.

THE ARMED FORCES—COAST GUARD

Coast Guard titles are the same as those in the Navy, except that the top rank is *Admiral*, *U.S.C.G.* indicates regular service, *U.S.C.G.R.* indicates the Reserve.

CHURCH DIGNITARIES/CATHOLIC FAITH

Personage	Envelope and Inside Address (Add City, State, Zip)	Formal Salutation	Informal Salutation	Formal Close	Informal Close	1. Spoken Address 2. Informal Introduction or Reference
The Pope	His Holiness, The Pope *or* His Holiness Pope Vatican City	Your Holiness: Most Holy Father.	*Always Formal*	Respectfully,	*Always Formal*	1. Your Holiness 2. Not introduced (His Holiness or The Pope)
Apostolic Delegate	His Excellency, The Most Reverend William R. Blank Archbishop of _____ The Apostolic Delegate	Your Excellency:	Dear Archbishop Blank:	Respectfully yours,	Sincerely yours,	1. Your Excellency 2. Not introduced (The Apostolic Delegate)
Cardinal in the United States	His Eminence, William Cardinal Blank Archbishop of New York	Your Eminence:	Dear Cardinal Blank.	Respectfully yours,	Sincerely yours,	1. Your Eminence *or less formally* Cardinal Blank 2. Not introduced (His Eminence or Cardinal Blank)
Bishop and Archbishop in the United States	The Most Reverend William R. Blank, D.D. Bishop (Archbishop) of Baltimore	Your Excellency:	Dear Bishop (Archbishop) Blank.	Respectfully yours,	Sincerely yours,	1. Bishop (Archbishop) Blank 2. Bishop (Archbishop) Blank
Bishop in England	The Right Reverend William R. Blank Bishop of Sussex (local address)	Right Reverend Sir:	Dear Bishop:	Respectfully yours,	Sincerely yours,	1. Bishop Blank 2. Bishop Blank
Abbot	The Right Reverend William R. Blank Abbot of Westmoreland Abbey	Dear Father Abbot:	Dear Father Blank:	Respectfully yours,	Sincerely yours,	1. Father Abbot 2. Father Blank

				Respectfully yours,	Sincerely yours,	
Canon	The Reverend William R. Blank, D.D. Canon of St. Patrick's Cathedral	Reverend Sir:	Dear Canon Blank:	Respectfully yours,	Sincerely yours,	1. Canon Blank 2. Canon Blank
Monsignor	Reverend Msgr. William R. Blank	Reverend Monsignor Blank:	Dear Monsignor Blank:	Respectfully yours,	Sincerely yours,	1. Monsignor Blank 2. Monsignor Blank
Superior of a Brotherhood and Priest[1]	The Very Reverend William R. Blank, M.M. Director	Dear Father Superior:	Dear Father Superior:	Respectfully yours,	Sincerely yours,	1. Father Blank 2. Father Blank
Priest	With scholastic degree: The Reverend William R. Blank, Ph.D. Georgetown University	Dear Dr. Blank:	Dear Dr. Blank:	Very truly yours,	Sincerely yours,	1. Doctor (Father) Blank 2. Doctor (Father) Blank
	Without scholastic degree: The Reverend William R. Blank St. Vincent's Church	Dear Father Blank:	Dear Father Blank:	Very truly yours,	Sincerely yours,	1. Father Blank 2. Father Blank
Brother	Brother John Blank 932 Maple Avenue	Dear Brother:	Dear Brother John:	Very truly yours,	Sincerely yours,	1. Brother John 2. Brother John
Mother Superior of a Sisterhood (Catholic or Protestant)	The Reverend Mother Superior, O.C.A. Convent of the Sacred Heart	Dear Reverend Mother: or Dear Mother Superior:	Dear Reverend Mother: or Dear Mother Superior:	Respectfully yours,	Sincerely yours,	1. Reverend Mother 2. Reverend Mother
Sister Superior	The Reverend Sister Superior (order, if used)[2] Convent of the Sacred Heart	Dear Sister Superior:	Dear Sister Superior:	Respectfully yours,	Sincerely yours,	1. Sister Blank or Sister St. Teresa 2. The Sister Superior or Sister Blank (Sister St. Teresa)

1. The address for the superior of a Brotherhood depends upon whether or not he is a priest or has a title other than superior. Consult the *Official Catholic Directory*.
2. The address of the superior of a Sisterhood depends upon the order to which she belongs. The abbreviation of the order is not always used. Consult the *Official Catholic Directory*.

CHURCH DIGNITARIES/CATHOLIC FAITH *continued*

Personage	Envelope and Inside Address (Add City, State, Zip)	Formal Salutation	Informal Salutation	Formal Close	Informal Close	1. Spoken Address / 2. Informal Introduction or Reference
Sister	Sister Mary Blank / St. John's High School	Dear Sister:	Dear Sister Mary:	Very truly yours,	Sincerely yours,	1. Sister Mary / 2. Sister Mary
Member of Community	Mother Mary Walker, **R.S.M.** / Convent of Mercy	Dear Mother Walker:	Dear Mother Walker:	Very truly yours,	Sincerely yours,	1. Mother Walker / 2. Mother Walker

CHURCH DIGNITARIES/JEWISH FAITH

Personage*	Envelope and Inside Address (Add City, State, Zip or City, Country)	Formal Salutation	Informal Salutation	Formal Close	Informal Close	1. Spoken Address / 2. Informal Introduction or Reference
Rabbi	*With scholastic degree:* Rabbi William R. Blank, Ph.D.	Sir:	Dear Rabbi Blank: *or* Dear Dr. Blank:	Very truly yours,	Sincerely yours,	1. Rabbi Blank *or* Dr. Blank / 2. Rabbi Blank *or* Dr. Blank
	Without scholastic degree: Rabbi William R. Blank	Sir:	Dear Rabbi Blank:	Very truly yours,	Sincerely yours,	1. Rabbi Blank / 2. Rabbi Blank

CHURCH DIGNITARIES/PROTESTANT FAITH

Archbishop (Anglican)	The Most Reverend Archbishop of Canterbury *or* The Most Reverend John Blank Archbishop of Canterbury	Your Grace:	Dear Archbishop Blank:	Respectfully yours,	Sincerely yours,	1. Your Grace 2. Not introduced (His Grace *or* The Archbishop)
Presiding Bishop of the Protestant Episcopal Church in America	The Most Reverend William R. Blank, D.D., L.L.D. Presiding Bishop of the Protestant Episcopal Church in America Northwick House	Most Reverend Sir:	Dear Bishop Blank:	Respectfully yours,	Sincerely yours,	1. Bishop Blank 2. Bishop Blank
Anglican Bishop	The Right Reverend The Lord Bishop of London	Right Reverend Sir:	My dear Bishop:	Respectfully yours,	Sincerely yours,	1. Bishop Blank 2. Bishop Blank
Methodist Bishop	The Very Reverend William R. Blank Methodist Bishop	Reverend Sir:	My dear Bishop:	Respectfully yours,	Sincerely yours,	1. Bishop Blank 2. Bishop Blank
Protestant Episcopal Bishop	The Right Reverend William R. Blank, D.D., LL.D. Bishop of Denver	Right Reverend Sir:	Dear Bishop Blank:	Respectfully yours,	Sincerely yours,	1. Bishop Blank 2. Bishop Blank
Archdeacon	The Venerable William R. Blank Archdeacon of Baltimore	Venerable Sir:	My dear Archdeacon:	Respectfully yours,	Sincerely yours,	1. Archdeacon Blank 2. Archdeacon Blank
Dean[1]	The Very Reverend William R. Blank, D.D. Dean of St. John's Cathedral	Very Reverend Sir:	Dear Dean Blank:	Respectfully yours,	Sincerely yours,	1. Dean Blank *or* Dr. Blank 2. Dean Blank *or* Dr. Blank

1. Applies only to the head of a Cathedral or of a Theological Seminary.

CHURCH DIGNITARIES/PROTESTANT FAITH *continued*

Personage	Envelope and Inside Address (Add City, State, Zip)	Formal Salutation	Informal Salutation	Formal Close	Informal Close	1. Spoken Address 2. Informal Introduction or Reference
Protestant Minister	*With scholastic degree:* The Reverend William R. Blank, D.D., Litt.D. *or* The Reverend Dr. William R. Blank	Dear Dr. Blank:	Dear Dr. Blank:	Very truly yours,	Sincerely yours,	1. Dr. Blank 2. Dr. Blank
	Without scholastic degree: The Reverend William R. Blank	Dear Mr. Blank:	Dear Mr. Blank	Very truly yours,	Sincerely yours,	1. Mr. Blank 2. Mr. Blank
Episcopal Priest *(High Church)*	*With scholastic degree:* The Reverend William R. Blank, D.D., Litt.D. All Saint's Cathedral *or* The Reverend Dr. William R. Blank	Dear Dr. Blank:	Dear Dr. Blank:	Very truly yours,	Sincerely yours,	1. Dr. Blank 2. Dr. Blank
	Without scholastic degree: The Reverend William R. Blank St. Paul's Church	Dear Mr. Blank: *or* Dear Father Blank:	Dear Mr. Blank: *or* Dear Father Blank:	Very truly yours,	Sincerely yours,	1. Father Blank *or* Mr. Blank 2. Father Blank *or* Mr. Blank

President of a College or University	*With a doctor's degree:* Dr. William R. Blank *or* William R. Blank, LL.D., Ph.D. President Amherst College	Sir:	Dear Dr. Blank:	Very truly yours,	Sincerely yours,	1. Dr. Blank 2. Dr. Blank
	Without a doctor's degree: Mr. William R. Blank President Columbia University	Sir:	Dear President Blank:	Very truly yours,	Sincerely yours,	1. Mr. Blank 2. Mr. Blank *or* Mr. Blank, President of the College
	Catholic priest: The Very Reverend William R. Blank, S.J., D.D., Ph.D. President Fordham University	Sir:	Dear Father Blank:	Very truly yours,	Sincerely yours,	1. Father Blank 2. Father Blank
University Chancellor	Dr. William R. Blank Chancellor University of Alabama	Sir:	Dear Dr. Blank:	Very truly yours,	Sincerely yours,	1. Dr. Blank 2. Dr. Blank
Dean or Assistant Dean of a College or Graduate School	Dean William R. Blank School of Law (If he holds a doctor's degree) Dr. William R. Blank Dean (Assistant Dean), School of Law University of Virginia	Dear Sir: *or* Dear Dean Blank:	Dear Dean Blank:	Very truly yours,	Sincerely yours,	1. Dean Blank 2. Dean Blank *or* Dr. Blank, the Dean (Assistant Dean) of the School of Law
Professor	Professor William R. Blank *or* (If he holds a doctor's degree) Dr. William R. Blank *or* William R. Blank, Ph.D. Yale University	Dear Sir: *or* Dear Professor (Dr.) Blank:	Dear Professor (Dr.) Blank:	Very truly yours,	Sincerely yours,	1. *Professor (Dr.) Blank* 2. *Professor (Dr.) Blank*

COLLEGE AND UNIVERSITY OFFICIALS *continued*

Personage	Envelope and Inside Address (Add City, State, Zip, or City, Country)	Formal Salutation	Informal Salutation	Formal Close	Informal Close	1. Spoken Address 2. Informal Introduction or Reference
Associate or Assistant Professor	Mr. William R. Blank *or* (If he holds a doctor's degree) Dr. William R. Blank *or* William R. Blank, Ph.D. Associate (Assistant) Professor Department of Romance Languages Williams College	Dear Sir: *or* Dear Professor (Dr.) Blank:	Dear Professor (Dr.) Blank:	Very truly yours,	Sincerely yours,	1. Professor (Dr.) Blank 2. Professor (Dr.) Blank
Instructor	Mr. William R. Blank *or* (If he holds a doctor's degree) Dr. William R. Blank *or* William R. Blank, Ph.D. Department of Economics University of California	Dear Sir: *or* Dear Mr. (Dr.) Blank:	Dear Mr. (Dr.) Blank:	Very truly yours,	Sincerely yours,	1. Mr. (Dr.) Blank 2. Mr. (Dr.) Blank
Chaplain of a College or University	The Reverend William R. Blank, D.D. Chaplain Trinity College *or* Chaplain William R. Blank Trinity College	Dear Chaplain Blank: *or* (If he holds a doctor's degree) Dear Dr. Blank:	Dear Chaplain (Dr.) Blank:	Very truly yours,	Sincerely yours,	1. Chaplain Blank 2. Chaplain Blank *or* Dr. Blank

UNITED NATIONS OFFICIALS[1]

Secretary General	His Excellency, William R. Blank Secretary General of the United Nations	Excellency:[2]	Dear Mr. Secretary General:	Very truly yours,	Sincerely yours,	1. Mr. Blank *or* Sir 2. The Secretary General of the United Nations *or* Mr. Blank
Under Secretary	The Honorable William R. Blank Under Secretary of the United Nations The Secretariat United Nations	Sir:	Dear Mr. Blank:	Very truly yours,	Sincerely yours,	1. Mr. Blank 2. Mr. Blank
Foreign Representative (with ambassadorial rank)	His Excellency, William R. Blank Representative of Spain to the United Nations	Excellency:	Dear Mr. Ambassador:	Very truly yours,	Sincerely yours,	1. Mr. Ambassador *or* Mr. Blank 2. Mr. Ambassador *or* The Representative of Spain to the United Nations (The Ambassador *or* Mr. Blank)
United States Representative (with ambassadorial rank)	The Honorable William R. Blank United States Representative to the United Nations	Sir: *or* Dear Mr. Ambassador:	Dear Mr. Ambassador:	Very truly yours,	Sincerely yours,	1. Mr. Ambassador *or* Mr. Blank 2. Mr. Ambassador *or* The United States Representative to the United Nations (The Ambassador *or* Mr. Blank)

1. The six principal branches through which the United Nations functions are The General Assembly, The Security Council, The Economic and Social Council, The Trusteeship Council, The International Court of Justice, and The Secretariat.

2. An American citizen should never be addressed as "Excellency."

BIBLIOGRAPHY

Prentice Hall's Business and Professional Division specializes in useful reference books for those in technical professions. The books provide a rich assortment, offering practical and essential data on a wide range of subject areas. The works related to effective written communications include:

Complete Book of Model Business Letters, Cresci

The Craft of Scientific Writing, Alley

Director's and Officer's Complete Letter Book, Van Duyn

Effective Business and Professional Letters, Sterkel

Handbook for Writing Technical Proposals that Win Contracts, Helgeson

Handbook of Business Letters, Frailey

Lifetime Encyclopedia of Letters, Meyer

Standard Book of Letter Writing, Watson